T0240849

Rubies and Implants

Bozena Arnold

Rubies and Implants

Aluminium Oxide and Its Diverse Facets

Springer

Bozena Arnold
Waldbronn, Germany

ISBN 978-3-662-66115-4   ISBN 978-3-662-66116-1   (eBook)
https://doi.org/10.1007/978-3-662-66116-1

This Springer imprint is published by the registered company Springer-Verlag GmbH, DE, part of Springer Nature.
The registered company address is: Heidelberger Platz 3, 14197 Berlin, Germany

Preface

All substances are made up of atoms. Every substance has a certain chemical composition. It determines the basic properties of the substance. In addition, the internal structure of the substance plays a very important role and exerts a strong influence on its properties.

In particular, it is very interesting when the composition and structure are the same, but in practice there are different materials. In this case, a substance has different faces. Such a substance with two faces is aluminium oxide – in nature a mineral and in technology a high-performance material. The famous diamond can also be counted among these substances.

The present book is a popular-scientific treatise on the alumina, on a single substance and very different materials – famous gems and modern ceramic materials. The former are old and long known, the latter are young and developed only recently.

My fondness for materials developed and became more and more pronounced during my many years as a professor of materials engineering, most recently at the HAW University of Applied Sciences in Hamburg. My admiration for minerals and their natural origin has developed to a special degree after my retirement. From these two preferences the idea for this book was born.

The book is aimed at all those who have basic chemical and technical knowledge and are interested in materials without seeking in-depth scientific knowledge. If, however, curiosity for further information should be aroused while reading, the book has fulfilled an important task.

Waldbronn, Germany   Bozena Arnold
October 2017

Contents

Fascinating Aluminium Oxide

Aluminium oxide occurs as a mineral in nature, is used as a high-performance material in technology and it is fascinating in every case. Why actually?

Aluminium oxide – that sounds dry at first, like any chemical term. What is supposed to be fascinating about it? Why does one feel like writing a book about it?

On the upper floor of the German Gemstone Museum in Idar-Oberstein, rubies are displayed in a showcase. One specimen (Fig. 1.1a) has an amazingly homogeneous colour, the special pigeon's blood colour. The gemstone sparkles, it is beautiful and it is admired.

Now we visit a textile factory. Several weaving machines are working loudly and intensively. The fibres run at high speeds through ceramic yarn guides. Thermal stress, friction, surface wear and, in the case of fibres, also electrostatic effects in production challenge these small components. The yarn guide (Fig. 1.1b) is colourless, works hard and can hardly be seen. Nobody admires him.

Two objects that could not be more different. And yet they are actually the same. Both are made of the same material, have the same chemical composition.

The gemstone and the ceramic thread guide are made of aluminium oxide. This simple chemical compound has two distinct worlds: a natural one, full of beauty and desire, and an artificial one, full of work and technical stakes. One world is magical, laden with myth and desire, bringing good luck and bad luck, promising wealth and beauty. The second world is technical, dry, sophisticated, cool and very applied. Isn't that fascinating?

We can find the aluminium oxide in the earth's crust and then we call it corundum or ruby or sapphire. We can also produce it on a large scale and then we usually call it alumina. Chemically, in all of these cases, it is a compound of aluminum and oxygen. The corundum and the gemstones ruby and sapphire are the natural manifestations of the aluminium oxide. The artificial alumina is mainly needed for the production of aluminium. However, it is also used as a material for valuable ceramic components, e.g. for thread guides in the textile industry and for implants in medical technology.

B. Arnold, *Rubies and Implants*, https://doi.org/10.1007/978-3-662-66116-1_1

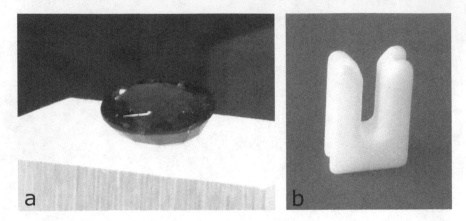

Fig. 1.1 Two aluminium oxides. (**a**) Gemstone (taken at the German Gemstone Museum in Idar-Oberstein), (**b**) thread guide

If these materials have the same chemical composition, why are they so different? That is a good question. You can find the answer in this book. Let's travel together through the worlds of aluminium oxide. Can this journey be interesting? Can we learn, learn, experience something along the way? Yes, certainly, because this material has a range of different properties such as high hardness, strong light refraction, good temperature and wear resistance. It is even said to have healing powers. Aluminum oxide is a true miracle material.

Our journey takes place in three steps. First we deal with the necessary information about the substance aluminium oxide. Then we travel through the world of gemstones and after that the world of alumina and technical ceramics.

The Aluminium Oxide Family

2

The head of the family represents a simple chemical compound of aluminum and oxygen with the formal name "dialuminum trioxide" and the trivial "aluminium oxide".

As already mentioned, this compound also bears well-known or even famous names such as corundum, alumina, ruby, sapphire. In any case, chemically it is the same material, namely the aluminum oxide. However, only chemically, as the origin, processing and application of these materials are different.

Let us first consider a small system (Fig. 2.1). Such a uniform representation is often used in the technical field.

Basically, the family consists of natural and artificial aluminium oxides. All materials mentioned in Fig. 2.1 have the same structure at the atomic level and thus have the same basic properties, such as high hardness or high melting temperature. However, they differ in:

- The origin: The natural aluminium oxide is found in the earth's crust, the artificial one is produced by us with the help of special processes.
- The macroscopic appearance: there are aluminium oxide crystals, i.e. solid bodies and aluminium oxide powder (often very fine). This influences the processing; with the powder we have to use different methods than with a solid crystal.

Corundum, the natural aluminum oxide, belongs to the minerals. About 4500 types of minerals are known, but only about 100 of them occur frequently. Corundum is a rock-forming mineral composed of the elements oxygen and aluminum. These elements are present in large quantities in the earth's crust. As a reminder, the Earth's crust is the outermost, solid shell of the Earth. It is on average 35 km thick, but in the internal structure of the Earth it is the thinnest shell.

Oxygen is by far the most common element in the earth's crust with a share of 46% and forms compounds with almost all chemical elements, e.g. various oxides with all metals. The light metal aluminium is involved in the structure of the earth's

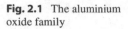

Fig. 2.1 The aluminium oxide family

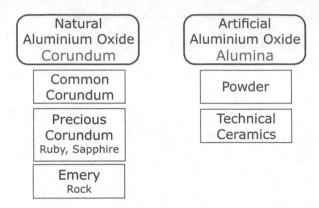

crust with about 8%, it is the third most common chemical element there and the most common metal.

Consequently, we should assume that there should actually be a lot of pure aluminium oxide (corundum) to be found in nature. But this is not the case. Mainly compounds of oxygen and silicon are present in the earth's crust: Silicon oxide (quartz) and many different silicates. Silicon is the second most abundant chemical element (28%) and is therefore available in much larger quantities than aluminium. The question of why so little corundum is formed in the Earth's crust will be dealt with in more detail in another chapter.

Let us briefly return to aluminium. In elemental form, the metal practically does not occur in the earth's crust. It is mainly a component of silicates such as feldspar and mica and thus of many rocks such as granite, gneiss, basalt or porphyry. In terms of quantity, aluminosilicates make up about 70% of rock-forming minerals. However, they are not used for the technical extraction of aluminium. Likewise, the natural aluminium oxide, corundum, is not used as an aluminium ore, but as an abrasive due to its high hardness. By far the most important aluminium ore is bauxite, which consists of aluminium hydroxides and mostly ferrous impurities.

Corundum is distinguished between the common and the noble colored corundum (ruby and sapphire). The fact that ruby and sapphire are actually only varieties of the inconspicuous corundum is something we can always marvel at. By the term variety we mean different formations of a mineral with regard to size, transparency and colour. Both varieties of corundum are known as precious gemstones and are models of colour. The ruby is a corundum coloured red by traces of chromium. The blue color of the sapphire is due to traces of titanium and iron.

In total, only seven gemstones are considered very valuable: Diamond, Aquamarine, Emerald, Topaz, Opal and the two colorful Corundum. In the gemstone world, we also speak of the classic "Big Four". This particular group consists of: Diamond, Emerald, and just Ruby and Sapphire. It is remarkable that these colorful aluminum oxides belong to this upscale gemstone elite.

Emery is a rock whose main component is small particles of corundum. We usually associate its name with grinding. And this is true, because ground emery serves as a good abrasive.

Alumina, the artificial aluminium oxide, is produced industrially or rather it is extracted. Various methods can be used for this purpose. The important impetus for the industrial production of the artificial aluminium oxide was the endeavour to produce metallic aluminium. The most important light metal today can only be produced on an industrial scale from suitable alumina. This requires many tonnes of the material, which cannot be recovered from the earth's crust. Natural pure aluminium oxide occurs far too rarely. The first stage in the production of aluminium is the production of the alumina, which is then in powder form and processed further. However, this powder also has other uses. With the help of sintering technology, we produce various ceramic components and products for technical and medical applications from alumina.

Polymorphism and Its Amazing Effect

3

Aluminium oxide is a polymorphic material. And what does that mean? The word "polymorph" stands for multiformity. In materials, we use it to describe the occurrence of a material in different modifications. These modifications have the same chemical composition, but differ in structure, in the spatial arrangement of the atoms and thus also in their properties. One composition and different properties – that's amazing.

In other words, the compound aluminium oxide has two faces, so to speak; it exists in two modifications. One we call α-(alfa)-aluminium oxide and the second γ-(gamma)-aluminium oxide. In both modifications, ionic bonding is found preferentially. The crucial difference between the two aluminum oxides is their internal structure, namely their crystalline structure. We will deal with the crystalline structure in the next chapter.

The α-aluminium oxide has a trigonal crystal lattice. This is our aluminium oxide, which occurs naturally as corundum, ruby and sapphire, as well as being produced in large quantities as alumina, which serves as a starting material for technical ceramics.

The γ-aluminium oxide has a cubic crystal lattice and is a colorless, loose, hygroscopic powder. This alumina modification does not occur in nature. Already at a temperature of about 800 °C, γ-aluminium oxide transforms into the α-modification. Because of this transformation, the γ-alumina is called active alumina. This aluminium oxide is an exceptionally porous material, the surface structure of which can be varied by the manufacturing temperature. Its porosity allows the γ-aluminium oxide to be used, for example, as a catalyst or catalyst support.

© The Author(s), under exclusive license to Springer-Verlag GmbH, DE, part of
Springer Nature 2022
B. Arnold, *Rubies and Implants*, https://doi.org/10.1007/978-3-662-66116-1_3

The two modifications of aluminium oxide are, after the two carbon modifications diamond and graphite, the most impressive examples of the fundamental correlation between the structure of a material and its properties. With the same composition, the two aluminium oxides (as well as the carbon modifications) nevertheless exhibit quite different properties.

From now on, we will devote ourselves entirely and only to α-(alfa)-aluminium oxide and omit the term "alfa".

The Inside of the Aluminium Oxide

4

Aluminium oxide has a crystalline structure inside it, whether it is of natural or artificial origin.

Crystalline means that the building blocks of a material, be they atoms or ions or molecules, are arranged spatially. Logically, this is always the solid state of the material. Crystals are described precisely and unambiguously geometrically. We speak here of crystal lattices and elementary cells.

Crystals rule the world. Yes, we can indeed say that, if we consider that most solid substances have a crystalline structure, for example, sand, steel, snowflake and also adrenaline in the solid state. We live on a crystal planet, in a world of crystals. It may sound pathetic, but also beautiful, as we associate the term "crystal" with beauty. Rubies and sapphires are a good example of this.

What kind of crystal lattice does our aluminium oxide have? (Remember: we are only interested in the α-(alfa)-modification). Depending on the source, we can read different things: The lattice is trigonal, rhombohedral, hexagonal, scalenohedral, rhombic. Which statement is then the correct one? The last information – rhombic lattice – is absolutely wrong, the others are all correct in some way.

Crystallography is a complicated science. It makes a classification of crystals, which is built up from 7 crystal systems, 32 crystal classes and 230 space groups. The crystal systems are distinguished according to the crystal axes and angles. Each crystal system can be assigned a crystallographic axis cross.

The hexagonal and trigonal systems use the same axial cross. For this reason they are sometimes combined and named as the hexagonal crystal family. This and the repetition of the name lead to confusion and to these different indications concerning the crystalline structure.

The difference between the hexagonal and the trigonal crystal system lies in the fact that the main axis in the hexagonal system (according to the name) is hexagonal, whereas in the trigonal system it is tridentate. Approximately 17% of all minerals belong to the two systems.

© The Author(s), under exclusive license to Springer-Verlag GmbH, DE, part of Springer Nature 2022
B. Arnold, *Rubies and Implants*, https://doi.org/10.1007/978-3-662-66116-1_4

In the trigonal crystal system (which is also called rhombohedral system) there is a class of crystals called scalenohedral. And to this belongs our aluminium oxide with all its varieties.

Thus, all the above designations concerning the crystal lattice of aluminium oxide contain a piece of truth. Quite correctly, we should say that aluminium oxide has a scalenohedral lattice belonging to the hexagonal crystal family and to the trigonal (rhombohedral) crystal system.

Let us now try to imagine the crystal lattice of aluminium oxide. It is shown schematically in Fig. 4.1. The lattice is made up of oxygen and aluminium ions.

The chemical formula of the aluminium oxide Al_2O_3 means that the quantity ratio of aluminium ions to oxygen ions in the lattice is 2:3. The electrical charge of the ions is different: aluminium ions have a triple positive charge, oxygen ions have a double negative charge. A crystal lattice should be electrically neutral.

The ions differ not only in their charge, they also differ, and significantly, in their size. The oxygen ions are about three times larger than the aluminium ions. The arrangement of the large oxygen ions determines the crystal lattice of aluminium oxide. The lattice planes formed by them are hexagonally stacked and densely packed. Six adjacent oxygen ions that touch each other form a cavity that geometrically represents an octahedral gap. The triple positively charged and much smaller aluminum ions are embedded in these gaps, so that each ion is surrounded by six oxygen ions. If the electrical charges are to balance out as required, there are only enough aluminium ions for two thirds of the gaps. One third of the gaps therefore remains unoccupied in a regular distribution.

The simple chemical composition and the gaps in the crystal lattice allow the substitution of some of the aluminium ions by ions of other metals, especially ions of chromium, vanadium, iron, titanium and manganese. Since these foreign elements can be incorporated in small quantities at most, they are called trace elements. These small changes hardly affect the properties of the aluminium oxide, apart from the colour. And so, with the help of foreign elements, the beautiful coloured varieties of aluminium oxide, ruby and sapphire, can be created.

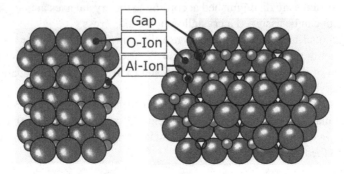

Fig. 4.1 Crystal structure of aluminium oxide – two sections. (Based on "New Materials", slide 22, Chemical Industry Fund)

So quite on the sidelines we can ask: How do the experts recognize a crystalline structure of materials? This is done using the fact that crystals cause diffraction of X-rays-, electron or neutron radiation. These diffractions can be detected and measured very accurately. The execution and evaluation of the diffraction experiments are rather complicated and one needs a complex apparatus. However, the results provide very valuable information and allow a detailed description of the crystals.

A Short Excursion into Etymology

5

Why is natural aluminium oxide called "corundum" or "ruby" or "sapphire"? What are the origins of these names? To explore this is the task of etymology, the study of the origin and origin of words and their meanings.

Etymology tells us that all three words have to do with colors. And this should not surprise us. The minerals mentioned are ancient and their names date back to a time when their composition was unknown. They were named according to their immediately visible properties. Of all the properties of a mineral, its color stands out the most. In second place may have been the hardness, that is, the resistance that the mineral opposes in practice, for example, when grinding.

The origin of the word "corundum" goes back to the ancient Indian languages Sanskrit and Tamil. In India, corundum was found at the beginning of historical times. The name can be derived from the Sanskrit word "kuruvinda" as well as from the Tamil "kuruntam". Both Indian words have a certain similarity and refer to the red color that the mineral can exhibit.

The names "ruby" and "sapphire" are used today as color names. The word ruby comes from Latin, where "ruber" means red. From the Aramaic "sappheiros" is translated as "the beautiful". However, until the Middle Ages, almost any blue rock or mineral was called "sapphire". It was not until chemical analysis revealed clear differences between various blue minerals. On the other hand, sapphires are not necessarily blue. They can occur in many other shades as well (Fig. 11.1). Today, however, we associate the term "sapphire" predominantly with blue gemstones.

And "aluminium oxide"? It is also a name. But for its origin we do not need etymology. The name was given according to the chemical rules and indicates the two elements that form the material: the aluminum and oxygen.

How Were Rubies and Sapphires Determined in the Past?

6

Even hundreds of years ago, people knew corundum and its colorful varieties. That is, they did not actually know the minerals. Neither in antiquity nor in the Middle Ages was the composition of corundum and of ruby and sapphire known and their determination often not possible. Thus, for a very long time other red stones were called rubies and blue stones were called sapphires.

To know what these stones really are, chemistry first had to become ripe for the task. Chemistry is a young science. Its beginnings go back only to the eighteenth century. The first great chemical achievement was Lavoisier's explanation of the process of combustion as the absorption of oxygen. After that, however, things really took off. Chemistry experienced an unprecedented development. There followed golden times for chemical analysis, for the discovery of new elements and for the determination of a wide variety of materials.

Nevertheless, still at the beginning of the eighteenth century it was believed that gemstones, which are characterized by particularly outstanding properties, also consist of particularly noble basic substances. A so-called precious earth was assumed to be the main component of all gemstones.

The famous German chemist Prof. Martin-Heinrich Klaproth (1743–1817) was the first to correctly determine the composition of colored corundum. He has, you might say, disenchanted rubies and sapphires.

In the Natural History Museum in Vienna, there is a replica of a chemical laboratory from around the end of 1790 (Fig. 6.1). Prof. Klaproth's laboratory probably looked similar.

Let's put ourselves in that time for a moment. One day in 1795, mineral samples were delivered from India. Among them was sapphire, which Klaproth describes in the following words: … *"That precious stone which we understand by the name of sapphire is distinguished by its blue colour, which is very pleasing to the eye, by its immense hardness and the exceptional brilliancy which arises from it when cut"* …

© The Author(s), under exclusive license to Springer-Verlag GmbH, DE, part of Springer Nature 2022
B. Arnold, *Rubies and Implants*, https://doi.org/10.1007/978-3-662-66116-1_6

Fig. 6.1 Replica of a chemical laboratory in the eighteenth century. (Image source: Sandstone CCBY 3.0)

Then the chemist began to conduct elaborate analyses of the oriental sapphire. After many steps of inorganic chemical analysis, which included grinding the precious sample as well as various digestion methods and filtration, Klaproth determined that sapphire consists of 98.5% alumina. He writes: ... *"we now find nothing in the digested sapphire but a simple aggregate of pure alumina"* ....

Was this result a disappointment? Certainly a surprise. But Klaproth also concluded correctly and full of admiration:... *"What a high degree of attraction and most intimate chemical connection must belong to it, and be at nature's disposal, in order to ennoble such a common substance as thunder clay into a natural body so distinguished by hardness, density, luster, resistance to the effects of acids, fire and weathering! So it is not the identity of the constituent parts alone, but the particular state of the chemical combination of these that determines the nature of the natural products formed from them."* ...

Thus, in 1795, the composition of sapphire had become known. Shortly thereafter, a classification of gemstones based on composition and crystal structure was introduced in England, in which ruby and sapphire were clearly named as varieties of corundum.

Those were turbulent times. Klaproth not only analysed gemstones, the discovery and naming of titanium is also one of his achievements. At the same time, progress was being made in other technical areas. In the 1790s, the French National Assembly adopted a new decimal system. A brass rod was cast to represent the new unit of length, the metre. In the same year, the new unit of grams was also defined as "the mass of a cube of edge length 1 cm of distilled water at a temperature of melting ice." As we know, both of these definitions no longer apply today. However, at the time it was a huge advance, as was the determination of ruby and sapphire. The first chemical analyses also opened the way to the production of synthetic gemstones, which took place immediately afterwards and about which we will learn more in this book.

The classical chemical analyses, in which, for example, acids act, are questionable and limited in the case of gemstones, because their application results in the destruction of the material. This must be avoided with such precious stones. It was only through modern mineralogy and analytical chemistry that the exact composition of minerals and gemstones was completely deciphered.

How Are Rubies and Sapphires Determined Today?

7

Today, gemstones are analyzed using appropriate physical methods. Gemstone analysis belongs to the field of gemology or gemology. It is a branch of mineralogy, which in turn is located in geology. Unlike geology, however, gemstone analysis does not involve taking a small sample, such as a piece of jewelry, and using it for analysis. All analyses are carried out using non-destructive methods. The gemstones can be reliably identified on the basis of their optical, physical, chemical and crystallographic properties.

Among the most important methods are optical examinations. Gemstones have many identifying features that are perceived by eye when we use optical devices such as the microscope, refractometer, polariscope and spectroscope as tools.

A simple to perform and at the same time very informative optical analysis technique is the determination of the refractive index. Every material has a specific value that depends on the type of material and the crystal structure. Gemstones with a high refractive index are usually particularly brilliant – for example, a diamond has a refractive index of about 2.4, that of corundum is only about 1.7. The refractive index can be measured with a refractometer.

Spectroscopic methods are based on a certain absorption behaviour of ions, especially in the visible spectral range. This results in certain absorption spectra that can be assigned to a mineral.

In addition to the optical behavior of gemstones in visible light, their observation in ultraviolet light of different wavelengths plays a role.

The physical examination methods are carried out with the aid of highly developed physical devices. These include methods using X-rays such as X-ray diffraction analysis or X-ray fluorescence analysis. Electron microscopy and electron beam microanalysis also provide valuable information.

A very modern method is Raman spectroscopy, in which a laser beam is used instead of an electron beam. It is focused in such a way that it causes the building

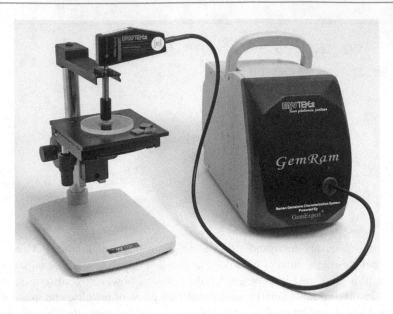

Fig. 7.1 Modern Raman spectroscope for gemstone analysis. (With the kind permission of GemExpert in Basel)

blocks (e.g. ions) of the sample under investigation to oscillate. This scatters the laser beam and forces it to change its wavelength in a structure-dependent manner. Finally, the chemical composition of the gemstone can be determined from this.

Since 2012, it has been possible to purchase a portable Raman spectroscope (Fig. 7.1), which can be used to analyse many types of gemstones quickly and reliably.

Corundum: The Natural Aluminium Oxide

8

We already know that corundum can be found in the earth's crust, where it was formed over a long period of time – not in large quantities, but quite frequently. A list of documented sites is long, too long to mention here. Actually corundum can be found on every continent. Places of discovery in Germany, Austria and Switzerland belong to it. Corundum has also been found in Greenland. Most often corundum forms long, columnar or prismatic crystals. These can sometimes be very large; corundum crystals about 1 m long and weighing up to 150 kg have been reported.

Since corundum is a mineral, we are interested in its optical properties. It can be transparent and opaque, although the transparent crystals are much rarer. Most often, corundum is cloudy, and then it cannot be used as a gemstone. Therefore, it makes sense to distinguish between the common and the precious corundum. The precious corundum is often compared with diamonds.

The precious corundum shines, the luster is strong and beautiful and is close in vividness to that of the diamond. Nevertheless, it is only the ordinary vitreous luster. The refraction of corundum is quite strong, but considerably less than that of diamonds. As a result, the colour dispersion, which we perceive as something beautiful, is low. The refractive indices of corundum differ only slightly for different colours (wavelengths). That is why the play of colours of the diamond can never be seen in corundum.

Don't know what refractive index and refraction are? Every material has a unique refractive index, which indicates how much light is refracted at the surface of the material, i.e. how far it is deflected from its original direction. The index also depends on the wavelength. The refraction of light is typical for all crystalline materials. Many of them also exhibit birefringence. Corundum is birefringent with an optical axis. However, its birefringence is low.

And now you are probably wondering: What is birefringence? Only amorphous materials and crystalline ones with a cubic crystal lattice have single refraction because their internal structure is the same in all directions. In all other materials, including corundum, the incident light beam is split into two beams (Fig. 8.1a). However, this phenomenon is only visible to the naked eye in a few materials. The birefringence can

B. Arnold, *Rubies and Implants*, https://doi.org/10.1007/978-3-662-66116-1_8

be nicely illustrated by a clear calcite crystal (Fig. 8.1b). When placed on a sheet of paper with writing on it, the writing appears double. If the crystal is rotated on the sheet, the position of one of the characters remains fixed, while the other moves back and forth. The fixed writing characterizes the "ordinary" beam, while the wandering writing characterizes the "extraordinary" beam (Fig. 8.1a). Unfortunately, the birefringence in the corundum is not so beautiful and clearly visible.

Another feature of many minerals is luminescence. Corundum also luminesces under the effect of supplied energy of a certain kind. The best known type of luminescence is fluorescence. Fluorescence is caused by UV radiation, among other things. In this process, the mineral is excited by the incident radiation – we call it the primary radiation. The primary radiation raises electrons to an excited state, from which they instantly fall back to the ground state, which is associated with emission of secondary radiation. The material fluoresces or glows in "dark light". As the primary radiation continues to fall, this process is repeated continuously. If we switch off the primary radiation source, the fluorescence also goes out. The fluorescence radiation has a longer wavelength than the primary radiation. In the case of corundum, the UV light causes flashing in shrill pink colours.

The density of corundum is about 4 g/cm³. From the point of view of materials engineering, the mineral still belongs to the group of light substances, which are counted up to the density of about 5 g/cm³. Among precious stones, however, corundum is one of the heaviest. Diamond, for example, is much lighter. Therefore, we can usually safely distinguish corundum from other similar-looking stones by weight. Liquid diiodomethane is best for this examination. Its density is exceptionally high at about 3.3 g/cm³. However, the corundum sinks rapidly to the bottom in the liquid, as it is even heavier.

Speaking of liquids – the melting temperature of corundum is very high at about 2100 °C. However, it is a theoretical figure, as natural aluminium oxide cannot be properly melted. When you hear about corundum produced by melting, note that then we are not talking about natural corundum, but an artificially produced one called electrocorundum, which is used as an abrasive.

The corundum breaks conchoidally, splintery and is only partially cleavable. If we rub it with cloth or leather, a positive charge is created which remains for a very long time. Corundum is not radioactive and cannot be magnetized.

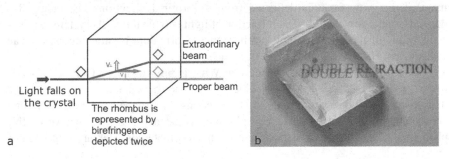

Fig. 8.1 Birefringence in crystals. (**a**) Principle, (**b**) Calcite. (Image source: commons.wiki-media.org)

We already know that there can be foreign ions in the crystal lattice of aluminium oxide. This can also occur in nature, and thus corundum becomes colored. This is how the famous gemstones ruby and sapphire are formed, varieties of corundum that are much rarer than common corundum.

Now let's move on to other properties of corundum and here we will also compare it to diamond.

The modulus of elasticity (Young's modulus) of corundum is large at 360 GPa, much larger than that of steel (210 GPa), but much smaller than that of diamond (approx. 1100 GPa). This important characteristic value is determined for solid materials in order to evaluate their elastic behaviour. By elasticity we mean the property of materials to reverse a deformation caused by external forces after the effect has been removed. The greater the modulus of elasticity, the greater the force required for elastic material deformation.

The thermal conductivity of corundum is low at 40 W/m K. Here corundum cannot compete at all with diamond, since the latter has the highest thermal conductivity of all materials with approx. 2100 W/m K. In the heat corundum expands very little, but more than diamond. However, we certainly would not have expected that the heat can strongly influence these minerals.

Chemical resistance of corundum is good and comparable to that of diamond.

We mention the most important property of corundum at the end. It is the hardness, which plays a major role in the gemstones. The corundum is the second hardest mineral after the diamond. Since the hardness is so important and also interesting, we devote the whole next chapter to it.

Fine-grained corundum is mined in high-yield emery deposits. In powder or paste form, it is used as an abrasive and polishing agent. In Fig. 8.2 we see sandpaper (emery paper) with corundum particles of two different grain sizes.

It is a well known application and we can assume that almost everyone has had corundum in hand as sandpaper. Corundum powder is now even used instead of quartz for sandblasting, as dangerous dust is produced when quartz sand is used.

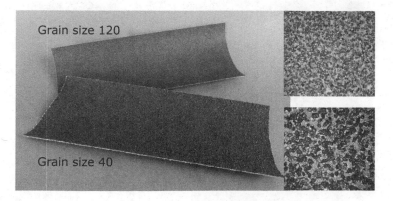

Fig. 8.2 Corundum abrasive papers for wood

How Hard Is Corundum?

Corundum and its varieties are very hard. After diamond, corundum is the second hardest mineral. Mineral well noted, but not the second hardest material. This will be briefly explained below.

In the mineral world, the hardness is traditionally given according to Mohs. Carl Friedrich Christian Mohs (1773–1839) was a German-Austrian mineralogist. He studied mathematics, physics, chemistry as well as mechanics and also worked as a mine foreman in the Harz Mountains. In 1812 Mohs was appointed professor of mineralogy at the Joanneum in Graz. During this time he developed the Mohs hardness scale, named after him. It was introduced into mineralogy in 1822 and is still a common method of hardness determination today. The method is simple. Mohs selected ten minerals of different hardness and arranged them so that each mineral is only able to score those that lie below it. Thus this scale has ten degrees of hardness; each degree of hardness being represented by a commonly occurring mineral. Here the first mineral (talc) is the softest, it can be scratched by all other minerals. We know talc mainly as powder, but it also exists as stone. Minerals up to hardness 2 can be scratched with a fingernail, those up to hardness 4 with a knife, and window glass can be scratched by all minerals with a hardness greater than 6. For comparison, the Mohs hardness of a Swiss steel knife is 7, which is about the same as the hardness of quartz. The tenth mineral in the scale is diamond. It is so hard that it cannot be scratched by any other mineral (or any other material!).

Scoring is also an excellent way to check the authenticity of a gemstone. Every gemstone has a hardness that is characteristic of it. If the stone can be scratched with a mineral of lower hardness, it is a fake.

The diamond has a hardness value of 10 in the Mohs scale and the corundum has a value of 9 (also its coloured varieties ruby and sapphire). Purely from the numerical value, the difference in hardness between the diamond and the corundum seems small. However, we must keep in mind that the steps of the Mohs scale are irregular. The degrees of hardness depend on the minerals chosen, so it is not an absolute scale, but one based on comparisons. As a result, the Mohs scale does not give a true

B. Arnold, *Rubies and Implants*, https://doi.org/10.1007/978-3-662-66116-1_9

sense of hardness. Corundum, which is natural aluminum oxide, is hard but much softer than diamond. To illustrate this even better, we will now look at another method of measuring hardness.

In technology, the Vickers method, among others, is used for hardness determination. Here, a square pyramid of hard diamond is pressed into a sample. A hardness value is calculated from the known force and the measured indentation. The method was developed for metals and is particularly suitable for them. However, it can also be applied to ceramic materials (thus including minerals), provided certain care is taken to prevent cracking. If we apply this Vickers method, we get the hardness value of 10,000 for the diamond and 2500 for the corundum. Now we can see quite clearly that the difference in hardness between the two minerals is considerable. We can see this fact well in Fig. 9.1, where the Mohs and Vickers hardnesses are plotted for the four hardest minerals. For comparison, the Vickers hardness of hardened steel and of the ceramic cutting material CBN (cubic boron nitride) is also shown.

When measuring hardness, we must always bear in mind – regardless of the measuring method – that the hardness of a crystalline material, including corundum, varies in individual directions. Hardness is an anisotropic property.

In the world of materials, the already mentioned cubic boron nitride – a synthetic ceramic – with the Vickers hardness of about 5000 is the second hardest material after the diamond. And the nitride is much harder than the corundum. Furthermore, other ceramics, such as tungsten carbide, which is found in the hard metals, have a higher hardness.

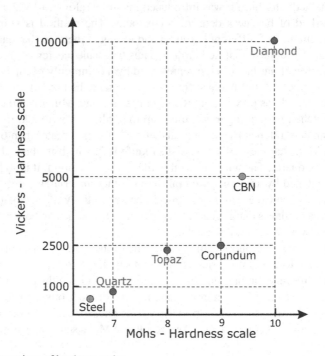

Fig. 9.1 Comparison of hardness scales

Also in the mineral world, strictly speaking, there is a mineral that is harder than corundum. It is called moissanite and it is the naturally occurring silicon carbide. Moissanite has a Mohs hardness of 9.5, so it actually ranks second in the hardness list after diamond. Moissanite is hardly found on earth (more likely on the moon). Since 1997, however, it has been produced synthetically as a diamond imitation and can challenge corundum for its second place.

Ruby: The Red Corundum

Ruby is the red variety of the mineral corundum, the natural oxide of aluminum.

Of all the gemstones known to man, ruby is probably the most popular. It is a preference that has been around for a long time. The ruby is the king of gemstones. Usually, we think that this designation belongs to the diamond. However, rubies are much rarer than diamonds and often they cost much more. Diamonds are found in the earth's crust, but rubies are not. Rubies with a magnifying glass are only available in a very small number and are hardly offered on the open market. This refers to larger clear crystals without visible inclusions, and not to the collections of very small grains that occur, for example, in the mineral zoisite (Fig. 10.1). These small ruby crystals look rather uninteresting.

Even larger cane rubies (Fig. 10.2a) hardly look any better; they appear dull and greasy. To make rough gemstones sparkle, they must be cut and polished (Fig. 10.2b). Then they show their great luster. Particularly clear red rubies are occasionally classified with the suffix "AAA." The similarity to the rating scale used by banks' rating agencies is astonishing.

The ruby is considered the stone of strong men. Among the men who were fascinated by rubies was Franz Stephan von Lohringen (1708–1765), the husband of Archduchess Maria Theresa. The emperor had keen scientific interests and wanted to fuse rubies by heating them with sunlight focused through lenses. Ignorant of what ruby actually is, he unfortunately achieved nothing. When heated, only a little gas (it was oxygen) was produced.

We already know that corundum, the natural aluminium oxide, shows fluorescence. This is also true for rubies, which fluoresce in UV light. This property can help identify the geographical origin of a stone. Rubies from Burma often fluoresce so intensely that the effect is noticeable even in sunlight; such stones literally seem to glow.

The color of ruby can be changed by certain treatments. The common methods are heat treatment, irradiation and oiling. However, this subject belongs to gemstone science and jewelry making.

B. Arnold, *Rubies and Implants*, https://doi.org/10.1007/978-3-662-66116-1_10

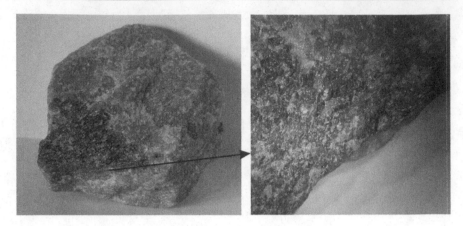

Fig. 10.1 Small ruby crystals in zoisite

Fig. 10.2 Ruby crystals. (**a**) Tubular rubies, (**b**) cut rubies. (Recorded in the German Gemstone Museum in Idar-Oberstein)

As the king of gemstones, the ruby has always been associated with numerous myths. The ancient Indians saw in him an imperishable inner fire, which could give a long life. It was said to bring luck in the game and to possess prophetic powers. When disaster threatened, it was said to darken its color. Rubies were worn by the Burmese as talismans to protect against illness, misfortune and injury. They were once known as "drops of blood from the heart of Mother Earth".

The ruby is already important in the Old Testament. In the bible, we can read that wisdom is more precious than rubies. You truly cannot put its value higher than that.

The aura of the ruby certainly also includes the fact that it is one of the healing stones. Healing with stones has an old tradition. One speaks of gemstone therapy or lithotherapy. The electromagnetic field of the body is said to absorb the radiation of the crystal. The recipient then becomes aware of the spiritual and emotional causes of his illness and recovers. However, there is no demonstrable effect of gemstones, only the idea of an effect.

Since when do we know rubies? It is believed that rubies were mined from pits in Burma during the Bronze Age. Rubies from Sri Lanka are said to have been known by the Greeks and Romans as early as 480 BC, making Sri Lanka one of the oldest known deposits. And where can we find rubies today?

The classic place of origin of rubies is considered to be the Mogok Valley (in German very appropriately "the ruby valley") in Myanmar (formerly Burma or Burma). This country is now considered the best origin for these stones in the world because of the best color and transparency of its rubies.

Otherwise, small rubies that are not good as gemstones can be found in many other countries. Rubies are found on all continents except Antarctica.

Almost everyone knows (ask your acquaintances) that ruby is a precious stone and that it is red. However, what is actually ruby, what is its composition – surprisingly, hardly anyone knows.

We have already read that it was not until the 1800s that the composition of ruby and sapphire was determined and their relationship recognized. And since 1835, rubies can even be literally made. How this works, you will learn in one of the next chapters.

Sapphire: The Blue Corundum

11

Why is the sky blue? Because the world stands on a giant sapphire and its color is reflected in the sky. A beautiful explanation, isn't it? That's what they used to think in Persia.

The most famous sapphire is native to England and has already played a role in two love stories. The first was in 1981, when Prince Charles presented a ring containing the 12-carat sapphire and 14 individual diamonds to the engagement of the future Princess Diana. In October 2010, the stone made its second appearance at Prince William's engagement to Kate Middleton. This piece of jewellery has given the sapphire star status in the gemstone world.

The name of sapphire is associated with the color, but the color of sapphires is not exclusively blue. Colourless, red-violet, yellow and even black sapphires are known. This variety of colours is clearly visible in Fig. 11.1.

Like rubies, rough sapphires look rather uninteresting (Fig. 11.1a). Only when worked into a gemstone does their beauty come to the fore.

When "sapphire" is mentioned, usually only the blue one is meant. In order to clarify the existence of other sapphire varieties, the addition "blue sapphire" has meanwhile become established. These different colors indicate that a sharp demarcation between the colored varieties of the corundum, the ruby and the sapphire, does not exist.

And where do we find sapphires? Like rubies, sapphires are found in many countries, but gem-quality stones are rare, though much more common than rubies. The "classic" source of Blue Sapphires is Sri Lanka (formerly Ceylon) – the oldest mining area ever mentioned. Sapphires can also be found in the famous Mogok Valley (Ruby Valley) in Myanmar.

Like ruby, sapphire has its aura and appears in mythological and religious stories. Blue sapphire is said to give its owner peace and joy. And the gemstone is also mentioned several times in the Bible.

B. Arnold, *Rubies and Implants*, https://doi.org/10.1007/978-3-662-66116-1_11

Fig. 11.1 Sapphire crystals. (**a**) Rough sapphire, (**b**) sapphires of different colours, (**c**) blue sapphires. (Recorded in the German Gemstone Museum in Idar-Oberstein)

Today, sapphires can also be artificially produced, or rather grown as crystals. Since sapphire crystals in particular find important technical applications, their industrial production is of great importance.

Why Is Ruby Red and Sapphire Blue? 12

Different corundum, ruby and sapphire differ only by their color, which is based on the presence of small amounts of foreign elements – so-called trace elements. For the consideration of corundums as minerals or as materials the color is unessential. For their use as gemstones, however, it has the utmost importance and significance.

How is a color created, why do we see a mineral, a stone colored?

Color, something that is a normality for us, something that we experience all the time, happens at a very deep atomic level of our world. Sunlight consists of several colors (i.e. wavelengths or frequencies) and is absorbed by substances in different ways. What remains is experienced by us, or more precisely by our optical system, the eye, as a particular color.

Ruby is red because it absorbs all other wavelengths (colors) of light. This is due to the chromium III ions contained in its crystal lattice. So far so good. But most often, these chromium ions cause substances (including minerals) to turn green. A prime example of this is emeralds, also well known gemstones. Thus, in the case of ruby, we are dealing with a peculiarity. Why do chromium ions cause the red colour of aluminium oxide?

Simplified, this effect can be explained as follows: In the crystal lattice of the aluminium oxide, the large chromium III ions occupy the places of the smaller aluminium III ions. These differences in size and the negative point charges on oxygen ions lead to a change in the energetic situation in the lattice. As a result, the transitions of electrons in the atom responsible for the color require more energy, more energetic, shorter-wavelength light is absorbed, and the ruby appears in the longer-wavelength red. Thus, it is not only the presence of foreign atoms that is responsible for the color, but also complex interactions at the atomic level. The color intensity is mainly influenced by the chromium content. Rubies with about 0.3% chromium have the preferred dove-blood red colour. If the oxidation state of the chromium changes, orange hues appear.

Ruby red is a color designation in its own right. Rubies are corundums that shine in very specific shades of red. But even the noblest ruby is only about 80% pure red, the rest are orange-, pink-, purple or violet tones.

In sapphire, too, foreign ions cause a change in the energetic state in the crystal lattice. The colouring ions, to which the blue colour of sapphires is due, are titanium and iron ions. The content of the trace elements and their interaction determine the colour appearance of the sapphire. If the iron content is comparatively low, for example, a green or orange hue is produced.

The crystal lattices of ruby and sapphire belong to the trigonal crystal system. This makes them birefringent and also pleochroic. This means that their color is different depending on the viewing direction. In the case of the trigonal system, there are two colors. Ruby shows a strong pleochroism from yellowish red to deep chimney red. In sapphires, on the other hand, this multicolourism is weaker.

One factor that decisively determines the color of a gemstone is not only light in general, but also the composition of the light. Of great importance is the intensity distribution of the wavelengths of a light source. A ruby dealer will always try to present his rubies in the light of a lamp that has a high red content. A sapphire dealer, on the other hand, would prefer neon light, which is characterized by a large blue component.

Why Are Only a Few Rubies and Sapphires Found?

If we consider that rubies, sapphires and other precious corundums are chemically nothing more than simple aluminium oxide, and that the two chemical components – aluminium and oxygen – are among the most common elements in the earth's crust, the question inevitably arises: Why are these minerals so rare? We are mainly concerned here with minerals of gemstone quality.

Aluminum and oxygen alone are not sufficient for the formation of precious corundum in the earth's crust. They must be present in sufficient quantities at the place of formation, but several additional factors are needed for corundum to crystallize at all. The key role for the formation – or rather for the "non-formation" – of ruby and sapphire is not played by aluminium, but by the ubiquitous silicon.

Wherever in nature the silicon (or the silicon oxide as quartz) comes too close to the aluminium, not the aluminium oxide will form, but one of the numerous aluminium-silicon minerals. From the point of view of corundum, this is far too often the case.

Thus, the first prerequisite for the formation of corundum is a low-silicon, ideally a silicon-free environment. In addition, it must also be rich in aluminium. This is not overly common in nature, but it is not extremely rare either. But, we are not interested in any corundum, we want to find colourful and large precious corundum. For this a relatively complicated environment is necessary, since beside the aluminum and oxygen still the color-giving elements, above all chrome, iron and titanium, must be present.

Furthermore, special geological conditions must always be present for the formation of corundum as well as ruby and sapphire. The occurrence of aluminium already gives an indication of a suitable source rock from which corundum or the two gemstones can be formed. Certain temperatures and pressures are also necessary, as well as a great deal of calm and continuity during crystallization, so that the crystals can develop. And that all this comes together happens exceptionally rarely in the earth's crust. Finding rubies and sapphires will forever remain a special event.

B. Arnold, *Rubies and Implants*, https://doi.org/10.1007/978-3-662-66116-1_13

In a Ruby Factory

14

We are in the middle of the Swiss Alps, in a place with a breathtaking view of snow-capped peaks. In one street there are inconspicuous factory buildings made of grey bricks. We go into a production hall. It is noisy and very warm. Rows and rows of similar looking jigs stand next to each other. There are hundreds of muffle furnaces with oxyhydrogen burners distributed over several blocks. Each gas burner is fed hydrogen and oxygen selectively and continuously. Small hammers strike the powder containers to release specific amounts of powdered aluminium oxide in a kind of "tapping concert" at a steady rate. The powder melts and accumulates in an orderly fashion to form a small rod. Depending on the addition of other metal oxides, rubies or sapphires are produced to order, so to speak, in the workshop. In just a few hours, materials are reproduced for which nature took millions of years.

At first glance it's hard to believe – a factory where rubies and sapphires are produced? Yes, exactly! The one briefly described above really exists.

Since Auguste Verneuil (1856–1913) published the results of his work, many such factories for the production of rubies and other precious stones have been established. The French chemist, son of a watchmaker, worked among other things for several years at the Natural History Museum in Paris. He dealt with many topics, but he owes his greatest fame to his invention of a commercial process for the production of artificial gemstones.

Verneuil experimented for 6 years until he achieved acceptable results. His most important achievement was a gas burner, which was operated with illuminating gas (gas mixture of hydrogen, methane, nitrogen and carbon monoxide) and oxygen and made it possible to generate high temperatures. Thus, nothing stood in the way of the synthesis of ruby and sapphire. Verneuil encrypted his notes and deposited them at the Paris Academy of Sciences. The public did not learn of this until 1902. Since rubies were always very expensive gems, the scientist's caution can be well understood. One could pathetically say that man's dream of artificially producing stones resembling natural gems was fulfilled at the beginning of the twentieth century.

© The Author(s), under exclusive license to Springer-Verlag GmbH, DE, part of
Springer Nature 2022
B. Arnold, *Rubies and Implants*, https://doi.org/10.1007/978-3-662-66116-1_14

In the history books of natural sciences, however, one can hardly read anything about the Verneuil method. It does not belong to the great discoveries. If we open the page for the year 1902, we find, for example, information about the theory of Rutherford and Soddy about the decay of atoms, in which one element transforms into another.

The Verneuil process is a flame melting process. This means that it is a growing process that does not require a crucible; the raw material is placed directly into a very hot flame. The resulting molten droplets precipitate onto a support and form a uniform crystal. A typical Verneuil apparatus is shown schematically in Fig. 14.1a. It is not particularly complicated and is composed of only a few parts.

In a container (1) is the raw material, the high-purity powdered alumina (we will learn later how this powder is produced). The purity of the powder plays a very important role – the purer, the better. A hammer (tapping mechanism) or an electromagnetic vibrator can serve as a material distributor. Its task is to distribute the aluminium oxide powder evenly in the flame. In a gas burner (2), hydrogen-oxygen mixture, i.e. oxyhydrogen gas, is burned. The gases (H_2 and O_2) are supplied by appropriate devices. Only with this explosive mixture can a necessary, very high temperature of approx. 2100 °C be reached. This much heat is necessary for melting aluminium oxide. On a refractory and rotating holder (3), the crystal grows layer by

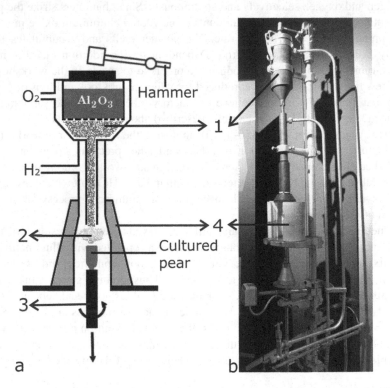

Fig. 14.1 Verneuil method. (**a**) Scheme. (Based on www.gemologyproject.com) (**b**) Model. (Recorded in the German Gemstone Museum in Idar-Oberstein) – Explanations in the text

layer on a crystal seed in the form of a so-called "culture pear". The growth takes place in a muffle furnace (4). As the crystal grows, it is slowly moved downwards with the aid of a lowering device so that it always remains in the ideal firing zone of the furnace. If about 2% chromium oxide is added to the starting material, then the sought-after red ruby crystals are produced. For the blue sapphire colour, 0.1–0.2% titanium oxide and some iron oxide are added to the aluminium oxide powder. Figure 14.1b shows a model of the Verneuil process, which can be seen in the German Gemstone Museum in Idar-Oberstein.

Synthetic corundum is mostly produced by the Verneuil method. However, today it is not the only one used for the production of gemstones. In general, these methods can be divided into three groups: Crystal growth from melt, crystal growth from solutions and crystal growth from gas phase (CVD method).

In addition to the Verneuil method described above, the Czochralski method also belongs to the first group. Here, the aluminium oxide powder and the colouring substances are inductively heated in a crucible. A rotating crystal nucleus is lowered from above onto the melt surface. During crystallization, the seed is pulled upwards without breaking contact with the melt surface. Ruby crystals for lasers are often produced in this way. The second group includes flux processes, which are also referred to as flux synthesis processes. In these processes, a substance (flux) is added to the doped aluminium oxide, which dissolves corundum and the colouring substances at relatively low temperatures (around 1000 °C). Oxides of lithium, lead and molybdenum are often used as the flux.

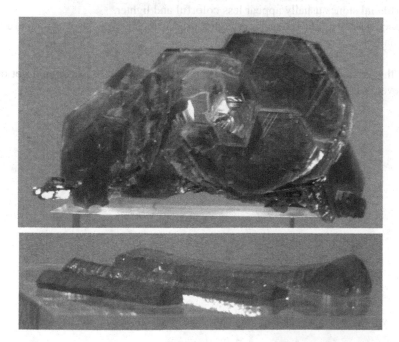

Fig. 14.2 Artificial rubies. (Recorded in the German Gemstone Museum in Idar-Oberstein)

A special type of flux method is the hydrothermal process. Here, water in combination with some salts is heated under pressure as flux. The artificial rubies shown in Fig. 14.2 were produced by this method.

The production of good artificial ruby and sapphire crystals requires a lot of time and care. Artificial gemstones are therefore anything but cheap. But natural gemstones still cost 10 to a 100 times more. However, artificial rubies and sapphires can be cut for jewelry and offered as an alternative to natural stones.

But more important is the wide technical application of artificial corundum. All methods allow to produce crystals with the desired properties and admixtures. Recently, artificial rubies and sapphires have become very important for modern technology. Today they can be found in many electronic and optical devices. Thus, future development of electronics depends not least on the development of crystal growing technology.

Finally, we ask an interesting question: can the natural rubies be distinguished from the artificial ones? It is not easy, because the physical-chemical properties are the same. Below you can read a small guide, which was compiled on the basis of recommendations from various gemstone experts:

- Look closely at the clarity of the ruby. Is it unnaturally clear? If so, it's probably artificial. Natural rubies often have visible inclusions.
- Look at the jewel under the microscope. If you see any bubbles, it is an artificial ruby.
- Look more closely at the color of the gemstone. Real rubies are intense deep red. Artificial stones usually appear less colorful and lighter.
- Touch the ruby. Real rubies feel rougher than artificial ones. Also, artificial rubies usually feel warmer.

With this advice, you may be able to recognize a real ruby if you should get one. However, it is best to contact a specialist laboratory.

The history of the laser began in May 1960 with a ruby. And since then, the laser has conquered the world. You probably also have a laser at home, for example in a DVD player.

The first solid-state laser was developed by the American Th. Maiman. For this he took a ruby crystal, not a natural one, but an artificial one. And this was not because of the price, but because of the composition and purity. Figure 15.1 shows a blank of an artificial ruby for laser application. The chromium content in a ruby suitable for laser must be only 0.02–0.05%, i.e. much lower than in natural rubies. In addition, the ruby rod must be very smooth. Its roughness should be about half the laser wavelength, so it must be only about 350 nm at the 694 nm wavelength. Polishing a ruby this smooth is a challenge. We remember that rubies are very hard.

The history of the invention of the laser, as with many other inventions, is littered with obstacles. Again, at first, no one believed that the laser would be used. This is almost unbelievable when we consider the importance of the laser today. Maiman studied engineering at the University of Colorado and then at Stanford University, where he earned a doctorate in physics in 1955. He first worked for Hughes Research Laboratories, a California aerospace manufacturer that was a hotbed of innovation at the time. In the beginning, Maiman developed a miniaturized version of the maser, a precursor to the laser. Then he wanted to concentrate light rather than microwaves. Due to discouraging reports from other research institutes, his superiors banned him from further laser research. However, his threat of dismissal and his private continued work on the laser caused him to relent, and he was granted money and an assistant. This enabled Maiman to develop the first working ruby laser. He initially submitted a description of his device to the journal "Physical Review Letters", which, however, refused to publish his manuscript. The equally high-ranking journal Nature accepted the manuscript and published it in August 1960, with the result that Maiman's invention was quickly replicated, with many modifications, by other researchers. Hughes Research Laboratories, on the other hand, still

© The Author(s), under exclusive license to Springer-Verlag GmbH, DE, part of
Springer Nature 2022
B. Arnold, *Rubies and Implants*, https://doi.org/10.1007/978-3-662-66116-1_15

Fig. 15.1   Blank of a laser ruby. (Recorded in the German Gemstone Museum in Idar-Oberstein)

gave insufficient support to the development of the laser. This reluctance, which seems hardly comprehensible today, was due to the fact that many researchers at the time were unable to identify any practical use for lasers. Maiman himself described the laser as "a solution in search of a problem". He received many honors for his research and also benefited from his invention himself by undergoing successful laser surgery in Munich in 2000.

Perhaps you would like to know how a ruby laser works? However, to understand how it works, we would have to go pretty deep into solid state physics. This would go beyond the scope of this little book. So we will leave it at that and trust physics. Interested readers can easily find appropriate descriptions – on the Internet or in relevant literature. Here is just some information in brief: A solid-state laser produces a monochromatic, parallel beam of light by exciting a suitable crystal, e.g. ruby, to glow by other beams. This can be done pulsed with a flash lamp or continuously, for example by another laser. In the process, electrons are raised to a higher energy level. They spontaneously fall back to the original energy level and emit light corresponding to a fixed wavelength. So the effect is to reverse the absorption that gives crystals their color. And as with the color of ruby, the trivalent chromium ions are responsible for creating the laser light in this case as well.

Today, the ruby laser has more of a historical significance. Its efficiency is comparatively low and the wavelength achievable by means of other lasers. Today, it is mainly used in dermatology. Its high pulse energy and the good absorption of its wavelength by melanin make the ruby laser suitable for the treatment of pigment spots and the removal of tattoos.

And what about the second colored corundum, the sapphire? Does it also have a significance for laser technology? Yes, it does, in fact, much more than the ruby. More specifically, it is a titanium:sapphire laser. (Its name is indeed spelled with a colon!) Like the ruby laser, it is also a solid-state laser and was first described in 1982 and technologically introduced in 1986. Its optical activity is brought about by fluorescence of titanium ions present as dopants in a sapphire crystal.

Most lasers have a fixed wavelength. What is special about the Titan:Sapphire laser is that it has an enormous bandwidth. Its wavelength is adjustable over a wide range from 670 to 1070 nm. This is due to the fact that several laser transitions are possible in this doped sapphire crystal. The titanium:sapphire laser dominates the range of ultrashort laser pulses. It is a femtosecond laser that can independently generate light pulses with an unimaginably short duration of about 100 femtoseconds (1 femtosecond = 10^{-15} s). For comparison: pulse duration of the widely used dye reader is in the range of picoseconds (10^{-12} s). The titanium:sapphire laser is mainly used in basic research and in laser medicine, e.g. for the correction of defective vision. In semiconductor technology, it is used for layer thickness measurements.

There is no question that laser technology owes a great deal to the two colorful varieties of aluminum oxide, ruby and sapphire.

Rubies, Sapphires and Watches

Perhaps the colorful varieties of corundum are in your wristwatch? If the watch is high quality, then with a high degree of certainty. In the watch industry, rubies and sapphires are used in applications that should not be underestimated. Let's take as an example the bearings that are necessary for moving parts in any mechanical watch. In a well-known way, stored potential energy is released in a controlled manner in mechanical watch movements and converted into a rotary movement of the display device (usually hand axes). The oscillations of a pendulum, a spring with a rotating pendulum (balance) or a crystal are used as a timing device. And in a clock, logically, there are many moving parts. Figure 16.1a shows the inside of a clock, actually only a section of it, and some moving parts are already visible there.

In small movements, precious stones are used as bearing stones for moving parts, since there is less friction between steel and stone than between two steel components. This reduces wear and tear and increases the precision of the movement by transmitting power evenly. A high-quality mechanical wristwatch with manual winding requires at least 15 functional jewels. In watchmaking, a *jewel* (French: *pierre*) is a bearing made of precious stones. Due to their perforation, jewels (also known as perforated jewels), together with the metallic pivots of the wheels or the balance, form sliding bearings for watches.

The first watch with a ruby-bearing balance was built in Switzerland in 1704. A few years later, the first watch completely equipped with ruby bearings was made. To this day, rubies with holes drilled through them are used as bearings. Their hardness facilitates sliding and greatly reduces wear, e.g. due to dust. At first, rubies of inferior quality were processed into bearing stones. After Vernueil's invention, which we already know, artificial rubies (and also sapphires) are now used. As we know, they are not imitations, they resemble the natural minerals in every respect. Compared to the natural ones, the artificial bearing stones even have advantages. Their perfect structure makes them extremely resistant to abrasion. They are pressed into the movement plate or screwed and adjusted within a gold setting.

B. Arnold, *Rubies and Implants*, https://doi.org/10.1007/978-3-662-66116-1_16

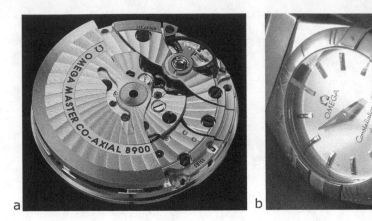

Fig. 16.1 Rubies and sapphires in watches. (**a**) Watch movement with ruby bearings. (With the kind permission of The SwatchGroup in Eschborn), (**b**) ladies' watch with sapphire glass

Have you heard of a sapphire crystal when buying a watch or wherever? The sapphire crystal is undoubtedly the king of watch glasses and with it a lot of aluminum oxide is consumed in the watch industry.

In the past, watches were equipped with normal window glass, but wristwatches needed a more shatterproof watch glass, so in the 1930s people switched to using acrylic glass, a plastic. Acrylic glass was not only lightweight, but more importantly, it was more shatterproof than regular glass and inexpensive. Unfortunately, plastic is susceptible to scratches, which is why mineral glass, i.e. tempered crystal glass, was increasingly used in the 1970s. However, mineral glass is only suitable to a limited extent for sports watches that are exposed to particular stresses, and so many manufacturers began to use the much more expensive, but also significantly more resistant sapphire glass instead from the 1980s onwards.

Today, scratch-resistant sapphire crystal is used in more and more watch models (Fig. 16.1b).

But now a problem arises: As a rule, a watch glass must not have a blue color, it must probably be colorless. Therefore, these glasses are not made of blue sapphire, but of colorless corundum. "Sapphire crystal" sounds much better, though, and there's actually no lying about it at all. We remember: red precious corundums are called rubies, all others we may call sapphires. Recently, the famous watch manufacturer Rolex also offers models with a unique green sapphire crystal. But usually watch glasses are colorless.

And there is something else that should occupy us briefly: Actually, the term "glass" is quite problematic. We use this word because it is a kind of window – we see the dial of a clock through it. However, in terms of materials, we only refer to glasses as amorphous materials (i.e. materials built up without a long-range order of the internal structure). Corundum is crystalline and therefore does not belong to glasses at all.

Now we have a good opportunity to talk about light transmission of materials. Have you ever wondered why we see everything geometrically unchanged through a window pane? This pane is made of glass, and because it is an amorphous material, light can pass through undisturbed. With crystalline materials it is different. There, light is refracted at the crystals, which greatly changes the image behind the material, sometimes to the point of being unrecognizable. This refraction of light is a disadvantage of sapphire glass. It results in stronger, disturbing reflections on the watch glass.

However, we see the dial intact with a sapphire crystal. Why? The answer is simple: sapphire crystals are anti-reflective. This anti-reflective coating, in turn, does not only bring advantages. The reduction of reflections (mirroring) on a watch glass is achieved by a vaporized chemical layer. This layer is nowhere near as resistant as the actual sapphire crystal. Some manufacturers make a reasonable compromise by only anti-reflecting the underside of the watch crystal. With this method, the anti-reflective coating is only 50% effective, but the resistant outer side of the sapphire crystal remains intact. On this occasion a tip: Sometimes it seems as if a sapphire crystal suddenly shows fine scratches. As a rule, however, these are only abrasions, i.e. marks left by softer materials on the glass. These marks can be easily removed with an eraser!

And again we remember that corundum and ruby and sapphire are very hard. In fact, a sapphire crystal can be scratched only by a diamond and materials such as tungsten carbide, silicon carbide and boron carbide. Practical experience has shown that sapphire glasses do not break more easily than mineral glasses, despite their greater hardness. However, you should not press on the sapphire crystal of your watch with very pointed objects.

Sapphire in Technology 17

At the same time sapphire is used much more often than ruby. There is a good reason for this: it is not red and can also be produced as a colourless material. Today, sapphire is the superior material for the most demanding optical applications. Firstly, because of its high transparency in the wide wavelength range and, secondly, because it is many times stronger than glass. The interesting optical properties in combination with the good chemical resistance and the high wear and temperature resistance make sapphire the leading material in optical sensors, spectroscopy, interferometry and other optical fields. A good example of this is sapphire lenses (Fig. 17.1), which are used whenever glass lenses cannot withstand the stresses. We find such lenses in medical endoscopes and in probes for high temperatures.

In scanners, sapphire is used as a scratch-resistant cover glass. It is also used in other applications where the highest wear resistance -is required: from bearing stones to cell phone displays to water/sand jets. Due to its unique combination of properties, sapphire can be exposed to the highest pressures and temperatures. With a melting point of over 2000 °C, it is an ideal material for high temperature applications, but is also used in cryogenics. We like to use sapphire as a material for the manufacture of refractory linings for blast furnaces or for the production of metal casting moulds.

Sapphire is resistant to most common acids and alkalis and this up to a temperature of approx. 1000 °C. Therefore, sapphire tubes, crucibles and optics in the chemical industry offer extremely long service lives compared to other materials or do not need to be replaced at all. Sapphire is replacing more and more quartz products due to its longer service life and lower tendency to contamination with good UV transmission at the same time.

In addition to its typical applications in insulation and thermal conduction, sapphire is widely used as an electronic substrate material due to its high and stable dielectric constant. Due to anisotropic properties (resulting from its crystalline structure) sapphire wafers form the basis for various products in semiconductor

© The Author(s), under exclusive license to Springer-Verlag GmbH, DE, part of Springer Nature 2022
B. Arnold, *Rubies and Implants*, https://doi.org/10.1007/978-3-662-66116-1_17

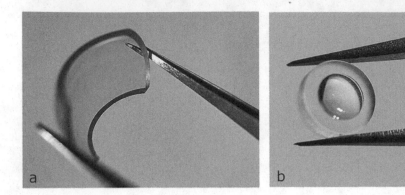

Fig. 17.1   Sapphire lenses. (**a**) Concave lens, (**b**) Convex lens. (With the kind permission of the company Badische Industrie-Edelstein BIEG in Elzach)

technology. Sapphire is currently the most widely used substrate material for gallium nitride-based LEDs (Light Emitting Diodes). This is a very modern and still young application of sapphire. Gallium nitride can be deposited epitaxially (we speak of epitaxy when at least one crystallographic orientation of the growing crystals corresponds to an orientation of the crystalline substrate) on sapphire, resulting in a blue light-emitting diode, which is popular.

As already mentioned, sapphire is a window material for medical and technical endoscopes. In addition to its biocompatibility, this is based on its high resistance in contact with biological tissue and medical fluids and its sterilizability in autoclaves. The use of sapphire scalpels or tips for laser surgery is common practice today. In the field of beauty & care dermatology, sapphire light guides form the contact between the high energy source and human skin, e.g. in the treatment of pigment disorders.

This list of application examples for sapphire could be extended. Certainly, the examples mentioned have convincingly shown that this special aluminium oxide is simply irreplaceable in today's technology.

Alumina: The Artificial Aluminum Oxide

The industrially produced artificial aluminum oxide we call alumina. However, its properties are the same as those of the natural oxide, corundum.

The term "alumina" is old and not entirely clear. Today, the name is used for various types of the aluminium oxide we produce artificially, also occasionally for aluminium oxide hydrates.

However, this name is often associated with minerals, especially clay and clay minerals. Actually, it is difficult to draw exact boundaries here. However, clay should not be treated in the same way as clay minerals or the clay (soil type), even if these contain a high proportion of clay. Other rocks also contain alumina, which is aluminium oxide.

Next to quartz, alumina is a very common component of minerals. Nevertheless, we cannot obtain pure aluminium oxide from it. It is often too strongly bound and thus not extractable.

A short excursion into the world of clay minerals helps us to better understand the difference between alumina and clay. In general, clays are chemical weathering products of silicates, in other words mixtures of aluminium and silicon oxides with water. Loam is clay contaminated by iron oxide and sand. The mixture of clay and water was once our first man-made material. Clay minerals include kaolinite, illite, and montmorillonite. You may never have heard of the last two names, yet the two minerals are very common. All three belong to the silicate group and contain aluminum. Kaolinite is also known as china clay. We could certainly make the alumina from these minerals. However, this is not considered economical and so we use another rock, bauxite, for this purpose.

Strictly speaking, alumina is understood to be the cubic γ-(gamma)-aluminium oxide. This oxide can be produced chemically. It is only stable up to 800 °C, then it transforms into the trigonal α-(alfa)-aluminium oxide. This transformation also happens during sintering (firing) of ceramics, which we will read about. Let us note the temperature mentioned. As we will learn, the artificial aluminium oxide is

B. Arnold, *Rubies and Implants*, https://doi.org/10.1007/978-3-662-66116-1_18

produced at much higher temperatures than 800 °C, and thus we are then always dealing with the α-(alfa)-modification.

As said before, the term "clay" should not be equated with "clay" or "clay minerals". However, it is often done that way, especially when writing generally about ceramics and their beginnings. The success story of ceramics is said to have begun more than 20,000 years ago by mixing clay with water. However, it was not the clay that is now produced on a large scale in alumina plants. It was various clay minerals that were found in the earth's crust and are still found today.

In 1795, the famous German chemist Klaproth (whom we have already met) called alumina a common substance. Yes, at that time metallic aluminum had no importance and the alumina was hardly used and appreciated.

Today it is quite different. Alumina is a very important raw material in various fields. Above all, the important light metal aluminium is obtained from the alumina in an electrolysis process. It is also used as a polishing agent for metals, catalyst and catalyst carrier in the chemical industry. Its use as an abrasive (electrocorundum) may be well known to some of you. A new and steadily growing application of alumina, aluminium oxide, is as a sought-after raw material for technical ceramics, in which case it is sometimes called "sintered alumina". For the cultivation of artificial rubies and sapphires we also need alumina, and of high purity. We need a lot of alumina. How is it produced today?

Bauxite and the Path to Artificial Aluminium Oxide

19

Until about 1880 we were hardly interested in the aluminium oxide. Except for beautiful and expensive jewellery made of ruby or sapphire, it found almost no application.

Verneuil's invention on the production of artificial rubies, and in particular the invention of an industrially applicable method for the production of metallic aluminium, almost abruptly aroused interest in aluminium oxide. The American Ch. M. Hall and the Frenchman P.L. Heroult, independently of each other, presented this new method in 1886. It was based on the reduction of aluminium oxide to aluminium by fused-salt electrolysis. As aluminium rapidly gained in importance, aluminium oxide also gained in importance at the same time. Now large quantities of the oxide were needed. But where was it to come from?

We know that there is a lot of aluminium in the earth's crust, but unfortunately not as aluminium oxide, but as aluminium silicates. Aluminium, after all the third most common element in the earth's crust, occurs everywhere, but never in pure form. Because of its strong tendency to react with non-metals – especially oxygen – it exists in nature only in compounds.

However, there are rocks that contain substances very related to aluminium oxide, namely aluminium oxide hydrates (aluminium hydroxides). And the rock called bauxite has a particularly large amount of these. From bauxite we can extract pure aluminium hydrate using the Bayer process (more about this later) and then from this pure aluminium oxide. To this day, bauxite is the most important source of aluminium oxide and therefore the most important ore containing aluminium, the most important raw material for aluminium production. The bauxite enables the transition to the modern technical application of the aluminium oxide.

The ore was discovered as early as 1821 by the French geologist Pierre Berthier and named after its place of discovery, Les-Baux-de-Provence in France. However,

© The Author(s), under exclusive license to Springer-Verlag GmbH, DE, part of
Springer Nature 2022
B. Arnold, *Rubies and Implants*, https://doi.org/10.1007/978-3-662-66116-1_19

the rock had to wait several more years for its application. Bauxite was formed over millions of years by chemical weathering of rocks containing aluminosilicate. It is composed of various minerals, mainly 55–65% aluminium oxide hydrates.

Substances that contain water are generally referred to as hydrates. In relation to metal oxides, such combinations with water are also called hydroxides. In our case, we accordingly have an aluminium hydroxide. The hydroxide ions of water are contained in the common crystal as negative lattice elements. The water is bound in the crystalline solid and thus it is appropriately referred to as crystal water. Both terms hydrate and hydroxide are often used as equivalent. We must come to terms with this.

Let us return to bauxite. It mainly consists of two hydrates: of trihydrate $Al_2 O_3 - 3H_2 O$ (or aluminium hydroxide $Al(OH)_3$) with the mineral name gibbsite and of monohydrate $Al_2 O_3 - H_2 O$ with the mineral name diaspor or boehmite. In addition, numerous admixtures occur, especially up to 25% iron oxide as well as silicon and titanium dioxide. If the iron oxide predominates, it is called red bauxite, if the silicon oxide predominates, it is called white bauxite.

For the most part, bauxite is extracted in opencast mines. Around 90% of bauxite deposits are concentrated in the countries of the tropical belt. The main mining areas are in Australia, China, Brazil, Guinea and Jamaica. Guinea has the largest bauxite reserves in the world. There are no large bauxite reserves in Europe. Australia alone currently supplies more than two-thirds of all bauxite. In Fig. 19.1 we can see bauxite lumps from the Weipa mine in Queensland, Australia.

Bauxite mining in Australia is favoured by the large deposits. Since the 2–10 m thick ore layers must first be freed of earth and cap rock before extraction, it is

Fig. 19.1 Bauxite lumps. (With the kind permission of Aluminium Oxid Stade)

necessary to attach great importance to the reclamation of the cap rock and earth material as early as the mining planning stage. The Australian aluminium industry is exemplary in this area.

If bauxite were composed only of aluminium oxide hydrates, we wouldn't have much of a problem extracting the oxide. The problem is the many admixtures. They have to be removed in order to obtain the purity of the aluminium oxide hydrate required for further processing. The Austrian Karl Josef Bayer came up with the idea of how to do this around 1890. The process named after him is still used today.

The Bayer Process: From Bauxite to Aluminum Hydroxide

To avoid any misunderstanding: The Bayer process has nothing to do with the world-famous chemical company that produces aspirin, among other things. It is the main production method of aluminium oxide, named after its – relatively unknown – inventor.

The raw material for the Bayer process is most commonly mined bauxite, a rock containing aluminium, which – as we already know – also contains many other substances (admixtures) in addition to aluminium oxide hydrates-, such as iron-, titanium and silicon oxides in particular. The main task of the process is to remove these admixtures.

Furthermore, for the sake of order, two processes should be kept apart:

- The Bayer process, i.e. the production of pure aluminium oxide hydrate (or more chemically formulated aluminium hydroxide) from bauxite,
- and the calcination process, i.e. the production of pure aluminium oxide from the aluminium oxide hydrate.

However, these two processes are often combined in books and the media and referred to as "the Bayer process". This second misunderstanding is difficult to dispel and can be explained by the fact that both processes usually take place in one plant, namely in an aluminium oxide plant. Figure 20.1 shows such a plant, which is located in Stade on the Elbe. In the photo, the storage area of the red bauxite is clearly visible.

The actual Bayer process is an alkaline circulation process and in chemical terms it is a wet digestion. In chemistry, digestion is generally understood to be a process in which an insoluble compound is converted into a soluble one. In the Bayer process, the basis for this is the different chemical solubility of the various oxides in caustic soda NaOH. Since the process is quite complicated, we will only mention the most important process steps here. The diagram of the process is shown in Fig. 20.2. The corresponding equipment in the aluminium oxide plant is marked in Fig. 20.1

Fig. 20.1 Aluminium oxide plant. (With the kind permission of the company Aluminium Oxid Stade)

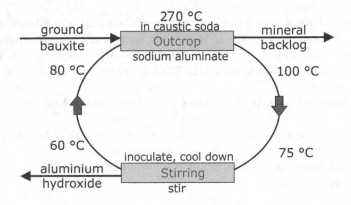

Fig. 20.2 Schematic of the actual Bayer process. (Adapted from www.aos-stade.de)

At the start of the process, the bauxite is finely ground and, with the addition of caustic soda, converted into a viscous suspension (chemists say: digested). The suspension is pumped through the pipes of a reactor and heated up in stages to approx. 270 °C. The lye dissolves almost specifically as the temperature rises and under pressure. As the temperature rises and under pressure, the lye almost specifically dissolves the aluminium compounds out of the bauxite. A new soluble aluminium compound called sodium aluminate is formed. It is cooled and thickened over several stages. This solution also contains solid and insoluble components such as iron and titanium dioxide. In large clarification tanks, these mineral residues are separated by sedimentation. The remaining residue is called red mud because of its intense red colour and is usually stored in a landfill.

After leaving the digestion plant, the sodium aluminate lye is highly supersaturated. Now this supersaturation must be removed. This is achieved by adding crystallisation nuclei (inoculation), by intensive stirring and by further cooling and pressure reduction. The precipitation of pure aluminium hydroxide begins, the so-called precipitation. The same reaction takes place as in leach digestion, but in the opposite direction. It is now a cyclic process. The difference to the initial situation is that now only aluminium hydroxide is present, without admixtures.

Thus we obtain the product of the actual Bayer process: the aluminium oxide trihydrate $Al_2O_3 - H_2O$, or aluminium hydroxide $Al(OH)_3$ – both terms and both formulae are chemically correct. In analogy to the use in practice, we will use both terms.

Aluminium hydroxide contains from 60% to 65% aluminium oxide, a small proportion of other oxides and the water of crystallisation bound in its lattice. Due to the chemical processes, the hydroxide is in powder form. It is a product in its own right and has a wide range of applications, for example in the manufacture of paper and glass, for cosmetic products or as a flame-retardant filler in carpets. However, the hydrate is not yet suitable for many applications, such as the production of technical ceramics, the cultivation of rubies or the production of metallic aluminium.

Let us return to the Bayer process. From a procedural point of view, we are very good at this method. However, we must not conceal the fact that this process causes real problems, especially for the environment. The biggest problem is the red mud I mentioned earlier. It contains many different and harmful substances, including numerous heavy metals (including mercury and arsenic). The characteristic red colour comes from iron (III) hydroxides. The red mud is hazardous waste. How much red mud is produced depends on the quality of the bauxite used. The highly alkaline sludge is usually landfilled or, in the worst case, discharged into rivers. Accidents in which a landfill dam breaks and the red mud can escape have unfortunately already occurred, e.g. in Hungary in 2010.

Despite all this, the Bayer process is indispensable for the production of alumina and consequently for the production of the most important light metal: aluminium.

The Forgotten Inventor

21

Only a few people know that the first process step in the industrial production of aluminium, which is used worldwide today, is an Austrian invention. It was the Austrian chemist Carl-Josef Bayer (1847–1904) who came up with the idea of using the wet digestion of bauxite with caustic soda for the production of aluminium oxide. He received two patents for his method in 1887 and 1892. Inspired by the invention of fused-salt electrolysis, in which metallic aluminium is produced from alumina, Bayer began to work on the new processes. The aluminium oxide, the alumina, was actually of little interest to him. Aluminium, the new and promising metal at the time, that was the incentive for his research.

Carl Josef Bayer was born in Belitz in 1847. His father was an architect and he only allowed him to study natural sciences after four semesters of architecture. After graduation Bayer worked as an assistant to the famous designer of the gas burner Robert W. Bunsen at the University of Heidelberg. In 1871 he received his doctorate, taught in Brno, and eventually went to St. Petersburg. There he married the niece of the then Russian prime minister, with whom he would later have six children, and invented the process named after him. Bayer then built several factories in various locations. In 1896 he finally settled in Recica ob Paki, now in Slovenia, where he built a laboratory and a villa – still preserved – that became a meeting place for scholars from all over the world. Unfortunately, an alumina factory was never built in Austria in which the bauxites found there could be processed. This factory was Bayer's greatest wish; he believed in the triumph of aluminium throughout the world (and he was not mistaken!). Unfortunately, Bayer did not live to see the great breakthrough of his process. The conversion of the first large plants to the Bayer process did not take place until after his sudden death in 1904.

Knowing today the enormous importance of aluminium and how fast the importance of aluminium oxide ceramics is growing, the Bayer process, which is more than 120 years old, was groundbreaking.

© The Author(s), under exclusive license to Springer-Verlag GmbH, DE, part of
Springer Nature 2022
B. Arnold, *Rubies and Implants*, https://doi.org/10.1007/978-3-662-66116-1_21

The brief recollection of the inventor is our final historical narrative in this book. Thus, we can now summarize what happened regarding aluminum oxide until the beginning of the twentieth century:

- 1774 – C. W. Scheele and J. Priestley recognize oxygen as an element when heating mercury oxide,
- 1795 – M. H. Klaproth determines that sapphire and ruby consist of alumina,
- 1821 – P. Bethier discovers the rock bauxite in France,
- 1827 – F. Wöhler obtains aluminium as an element by reduction of aluminium chloride with potassium,
- 1886 – Ch. M. Hall and P. L. Heroult invent the electrolytic production of aluminium from aluminium oxide (alumina),
- 1887 and 1892 – C. J. Bayer registers his patents for the process of producing alumina hydrate from bauxite.

It must be mentioned on this occasion: The nineteenth century was unprecedentedly rich in discovery. There has never been anything comparable since.

Calcination: From Aluminium Hydroxide to Aluminium Oxide

22

After the historical detour, we come back to the production of the artificial aluminium oxide, the alumina. We had just reached the end of the Bayer process and first produced the aluminium oxide hydrate (or aluminium hydroxide).

This brings us to the second process, calcination, in which pure (i.e. dehydrated) aluminium oxide is produced from the aluminium hydroxide.

First, let's take a close look at the chemical formula of aluminum hydroxide: $Al(OH)_3$. Recognize its similarity to the chemical formula of aluminium oxide Al_2O_3? If we take two molecules of hydroxide and remove three molecules of water from it, we are left with one molecule of aluminium oxide. This basic chemical principle is shown in Fig. 22.1.

What works for two molecules also works – with the help of suitable technology – for, say, two tonnes of aluminium hydroxide. And how can the water be removed technologically? By heating, isn't it?

Following this idea, the aluminium hydroxide is heated to approx. 1000 °C in a special furnace during calcination. In the fluidised bed process, the water of crystallisation is removed from the hydroxide. The result is aluminium oxide, which is available as a white powder. However, it still contains – usually very small – residues of foreign oxides, such as silicon-, iron-, sodium-, magnesium and calcium oxide. Their proportions determine the degree of purity and the subsequent use of the aluminium oxide. Aluminium oxides with low sodium oxide contents are particularly valued. The grades produced by calcination usually contain 98.5% aluminium oxide.

Almost 90% of the artificial aluminium oxide has a short life span, because this quantity is consumed for the production of metallic aluminium. This is what happens when a substance has a higher, more important purpose, so to speak. Aluminium is the second most important metal in the world economy and -industry. At this point we could ask a so-called stupid question: How do you get from this powdery aluminium oxide to metallic aluminum? This task belongs to another branch of

B. Arnold, *Rubies and Implants*, https://doi.org/10.1007/978-3-662-66116-1_22

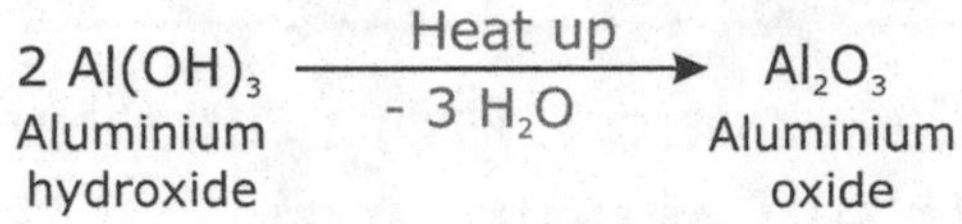

Fig. 22.1 Basic chemical principle of calcination

$$2\ Al(OH)_3 \xrightarrow[-\ 3\ H_2O]{\text{Heat up}} Al_2O_3$$

Aluminium hydroxide   Aluminium oxide

chemistry, namely electrochemistry, because here we have to use electrolysis. But that's a whole other story that has little to do with the subject of this book. The answer to the question can be found quickly on the Internet and in relevant literature.

We are interested in the remaining amount of aluminium oxide that is still available and is divided into different grades. These grades are called alumina or calcined corundum and come on the market preferably in finely ground, also super finely ground. We need the aluminium oxide, for example, in the manufacture of glass or for refractory linings. However, we are particularly interested in the application of aluminium oxide as technical ceramics. Now we ask again: what can be done with a powdered product? A great deal, as we will soon learn.

The World of Ceramics

23

Aluminium oxide is the most important ceramic material today and will remain so for some time to come. Before we delve further into aluminium oxide, some general information about ceramic materials is essential.

The term "ceramics" cannot be clearly defined. However, it is well known. Each of us can easily name a ceramic object and also show one.

Experts say about ceramic materials that they belong to the non-metallic inorganic materials. Yes, they are not metals, and neither are materials based on carbon compounds. The latter we call non-metallic organic materials or plastics (or colloquially plastic).

Ceramics are all inorganic, non-metallic materials and products that are manufactured using powder technology (sintering technology) and only acquire their final properties through a high-temperature process. The production process comprises the following steps: production of the starting powder, mixing of different powders, usually in a suspension medium, shaping by various methods, in any case firing or sintering and usually also finishing. Due to the manufacturing process, ceramics are polycrystalline materials, whereby fine glass phases or pores can also be found between the grains, which are only a few micrometers in size.

Ceramic materials are now considered to be very modern materials. It is hard to believe that they were the first materials that people consciously used. First we used stones as materials, from which mainly tools were made. Accordingly, we call this first epoch the Stone Age. Stones are non-metallic and inorganic, so they are a special type of ceramic.

The history of true pottery began more than 20,000 years ago, in the Paleolithic Age, when people discovered that clay mixed with water (we've already talked about the easy-to-find rock) could be kneaded and shaped. And that these fragile clay objects become more durable and harder when placed in fire. We still use this process principle today and call it the sintering technique. More about this in the next chapter.

© The Author(s), under exclusive license to Springer-Verlag GmbH, DE, part of Springer Nature 2022
B. Arnold, *Rubies and Implants*, https://doi.org/10.1007/978-3-662-66116-1_23

And so ceramic materials have been conquering the world for thousands of years. Around 6250 BC, potters in what is now Turkey produced ceramic spindles for weavers. Three thousand years later, ceramic sewer pipes provided regulated irrigation and drainage for the first time in Habuba, in what is now Syria. These ceramic products were a huge advance for mankind.

However, the history of aluminium oxide ceramics is not so old. It is actually a very young material. Its beginnings date back to the first decades of the twentieth century, which is logically related to the development of its manufacturing method just described. The first mention of aluminium oxide ceramic as a material for spark plug insulators is in a patent specification from 1928. Even today, these insulators are made from the oxide.

Aluminium oxide belongs to the high-performance ceramics, which is constantly used in new areas and contributes to technical development. The required raw materials are almost exclusively produced artificially and components and objects are manufactured from them using sintering technology. This manufacturing method, like the ceramic materials themselves, is on the one hand old and – today further developed – very modern and future-oriented.

Useful Sintering Technology

24

Let's imagine we are at a company that produces ceramic products from aluminium oxide. How are they made? As we already know, the raw material is in powder form, this results from its production method. For powdery raw materials we use the sintering technique. Here, since powder is not yet a real material, the material and a product are created in a single process. Several process steps are necessary for this, which can only be briefly discussed here. Detailed information can be found in the further literature. The main process steps in the sintering of aluminium oxide ceramics can be followed with the aid of Fig. 24.1.

First, the powdery aluminium oxide is mixed with plasticizers such as clay, kaolin or even magnesium compounds as well as sintering aids (to reduce porosity). This determines the composition of the ceramic material.

The component can be formed under pressure in special tools via axial or isostatic pressing or pressure-free as slip casting in suitable moulds. In this process, the powder particles are compacted and formed into a coherent shape. The so-called green or white bodies are then mechanically processed. Due to the high hardness of sintered aluminium oxide, mechanical processing such as sawing, grinding, turning or drilling must be carried out in the so-called green state, even before sintering. After sintering, only hard machining is possible, with tools made of diamond or cubic boron nitride (CBN). These materials are harder than aluminium oxide.

Sintering is a heat treatment in which a compact material is created from the powder mass. At temperatures of around 1500–1700 °C, further compaction and solidification of the green or white bodies occurs. No change in shape takes place, but shrinkage, a change in volume of the workpiece, often occurs. When the powder particles are compacted, pores are left behind which can take on a function. The porosity can be controlled. Post-treatments or finishes are usually necessary and are selected depending on the material and the use of the components.

The processes involved in sintering are complex. In general, sintering is based on diffusion processes. The driving force is the reduction of the surface energy. The

Fig. 24.1 Process steps of
the sintering technique

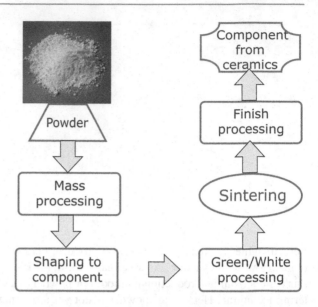

process depends on the purity and grain size of the starting powders, the sintering temperature and the type of ambient medium. Basically, we distinguish between the following processes: Solid phase sintering, liquid phase sintering and reaction sintering.

The result of the sintering process is adjusted by a defined temperature-pressure-atmosphere-time curve. The pre-compaction of the blank, the particle sizes and their distribution and, last but not least, the chemical composition of the raw material also influence the result.

In practice, sintering is often also referred to as firing. However, we should actually speak of sintering when the physico-chemical reactions are the focus of consideration, and of firing when it is primarily a matter of the actual furnace operation.

Aluminium oxide ceramics usually have a homogeneous structure with grain sizes of 10–30 μm. This granular structure is clearly visible under a scanning electron microscope (Fig. 24.2).

The microstructure shown in Fig. 24.2 consists of grains, the majority of which are smaller than 10 μm. There are both coarse-grained and fine-grained ceramics. By reducing the grain size to 1–3 μm, a significant increase in strength can be achieved.

With the appropriate selection of raw materials, i. e. qualities of aluminium oxide produced in the aluminium oxide plant, and the process parameters, dense, homogeneous products with desired properties can be produced.

If we want to buy aluminium oxide for ceramics production, we have a choice between four groups of alumina, which differ in the oxide content: the lowest group (called C 780) contains 80–86% aluminium oxide, and the highest group (C 799) means an oxide content greater than 99%.

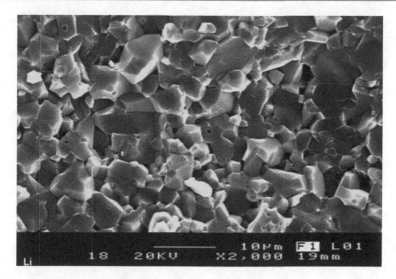

Fig. 24.2   SEM image of an aluminium oxide ceramic

Ceramic components made of aluminium oxide are produced exclusively by sintering. But why actually? Could one perhaps use a simpler method based on the solidification of a melt? Unfortunately, this is not possible because the aluminium oxide has a very high melting temperature. We would already have problems with a crucible suitable for this. With such high-melting materials, only the sintering technique makes sense and is successful.

Is Aluminium Oxide Ceramic Loadable?

25

This is a good question, especially since we usually perceive ceramics as a brittle material. And the answer is: the aluminium oxide ceramic is loadable, its strength characteristics are good. At the same time, however, this answer is insufficient and even misleading, because in the case of strength, it is essential to consider the nature of the load. How does the force act: in tension or in bending or in compression? Accordingly, we name the characteristic values as the tensile strength, the flexural strength and the compressive strength. Is it an abrupt, rapid load or a slow, quasi-static load?

Before we go any further into strength, we must again mention this very bad property of aluminium oxide, of ceramic materials in general: They are brittle, or to put it in technical terms, they have a very low toughness. This means that stress peaks that occur in the material or even a sharp change in temperature can lead to fracture. In other words, ceramics are sensitive to impact loading. The tendency to brittleness is indicated by a special characteristic value. It is called fracture toughness and its determination is rather complex. The low value of fracture toughness of only about 3.5 MPa $\sqrt{m}$ for pure aluminium oxide proves that it is very brittle.

The most important type of load for determining the strength of materials is the tensile load. However, with tensile loads we must note that they are unfavorable for ceramic materials. To understand this, we should remember the chemical structure of aluminium oxide. Ionic bonding is responsible for its cohesion. Primarily heteropolar bonds exist between the oxygen anions and aluminum cations. Thus, in contrast to metals, the electrons are firmly positioned. If a mechanical force acts, no sliding processes can be triggered. If the external mechanical forces exceed the bonding forces between the ions, the ceramic material breaks. However, since these binding forces can reach very high values, we are initially surprised that aluminium oxide breaks even at relatively low tensile stress. What is the reason for this? Internal defects play an important role here. They occur inside the material during powder production and on the surface through use. These are very short straight or curved as well as flat cracks. These cracks are pulled apart by tensile stresses, and a stress concentration, a stress peak, occurs at the ends of the crack. This leads to a local exceeding of the strength. The process builds up until the remaining stress-transmitting area becomes too small and the ceramic product breaks spontaneously and brittly.

Due to this behaviour, the flexural strength is almost exclusively determined and specified for ceramic materials instead of the tensile strength from the tensile test. Additional reasons for the preference of testing under bending load are the simple specimen geometry, the simple production of the specimens and their unproblematic clamping during the test.

In the case of ceramics, the so-called four-point bending test is most frequently performed (Fig. 25.1). In the case of four-point bending, the largest bending moment M_{max} remains constant between the two force application points. The four-point bending strength of aluminium oxide ceramics of typical grain size of about 20 μm is about 200 MPa at low purity levels and about 300 MPa at high purity levels.

The modulus of elasticity (Young's modulus) can also be determined in the bending test. This characteristic value of aluminium oxide ceramics is high, in the range of 200–300 GPa. The strong bonding forces are responsible for this. The modulus of elasticity is the basic indication of the stiffness and stability of structures, which is why it is particularly important in design. For comparison: steel – by far the most important construction material – has a modulus of elasticity of about 200 GPa. And that is good enough to realize bold steel constructions. Aluminium oxide looks even better if we consider specific properties, i.e. properties related to density. Then its specific modulus of elasticity is almost three times higher than that of steel.

Compared to other materials, the compressive strength of aluminium oxide is particularly high, its characteristic values are in the range of 2000–4000 MPa. It is typical for ceramics that the compressive strength is much higher than the strengths

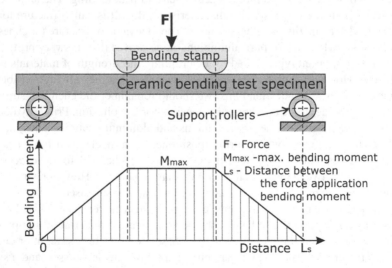

Fig. 25.1 Principle of the four-point bending test. (Based on "Mechanical engineering materials technology", Europa Lehrmittel, 2007)

under other types of loading. When compressive stress is applied instead of tensile or bending stress, the cracks already mentioned are pressed against each other. A frictional connection is created and the course of stress in a ceramic material is unhindered.

In practice, it is important how materials behave under load at high temperatures (above approx. 500 °C) over a longer period of time. We are interested in the creep rupture strength. Due to impurities caused by the necessary sintering aids, the aluminium oxide tends to creep and thus to lose strength.

The specification of strength values of ceramic materials is almost always subject to uncertainty, which is due to the relatively large scatter of the values. The reason for this is again to be found in the sintering technique. The manufacture of products from powder inevitably causes material inhomogeneities or volume and surface defects which have a size and orientation distribution. Even dense and non-porous aluminium oxide ceramics show a strength scatter of ±20%. It is therefore necessary to test a larger number of specimens in order to take this scatter into account.

In these strength-related representations, we should keep in mind that aluminium oxide is not a structural material and thus its good mechanical properties can be put to little practical use.

The determination of characteristic strength values is always associated with the destruction of the respective sample. These are extreme tests in which we reach the limit of the load-bearing capacity of a material. However, in the case of artificial aluminium oxide, in the case of aluminium oxide ceramics, the above-mentioned mechanical-technological tests are carried out regularly. And what about natural aluminium oxides, ruby and sapphire? Well, destroying samples from the gemstones is certainly disliked. And since they are not really used as construction materials, we are hardly interested in their mechanical strength parameters.

Advantages of Aluminium Oxide Ceramics

26

In addition to interesting mechanical properties, pure aluminium oxide ceramics, as an optimally sintered material with a dense homogeneous microstructure, has further advantageous sides.

With its density of approx. 3.6 g/cm^3, ceramics, like natural aluminium oxide, is a lightweight material. This is due to the light atoms of oxygen and aluminium of which it is composed. We remember: Light materials are those whose density is below 5 g/cm^3 . Due to its low density, aluminium oxide has good specific, i.e. weight-related, properties. This consideration has already been mentioned as important for mechanical properties. However, aluminium oxide ceramics are preferably used as a functional material. In such applications, the density and weight of components are not of primary importance.

As expected, the hardness of aluminium oxide ceramic of about 2000 Vickers units is very high. Just as high as the hardness of the corundum, as there is no chemical difference between the two materials. The corundum is the second hardest among the minerals. However, aluminium oxide ceramics do not take second place among ceramics. Many ceramic materials are even harder, for example, the cubic boron nitride or tungsten carbide. It is good that the high hardness of the aluminium oxide ceramic remains almost constant over a wide temperature range.

The high hardness is associated with the excellent wear resistance of the aluminium oxide. Wear resistance is a system variable, as it depends on many factors, and cannot be described with a single key figure. Compared to metals, aluminium oxide ceramic is 50–100 times more wear resistant. This excellent wear resistance is the reason why we need this material. It is decisive for technical applications of aluminium oxide ceramics.

However, the high hardness also has a negative side. Hard materials are brittle, so is the aluminum oxide. Its brittleness is something we absolutely have to take into account in practice. It often leads to catastrophic fractures. This is unfavorable and means that ceramic components must be handled carefully. Like, for example, they must not fall on hard ground.

Aluminium oxide has a very high melting temperature of approx. 2050 °C. As a result, ceramic components, as we know, have to be manufactured using sintering technology. However, the high melting temperature also means that the aluminium oxide has a very good temperature resistance. It can be used permanently, at any rate without mechanical stress, up to approx. 1500 °C. With certain grades, its use is even possible up to approx. 1900 °C.

The application temperature of aluminium oxide is much higher than that of the heat-resistant steels or the famous nickel-based superalloys. However, temperature changes do not tolerate aluminium oxide ceramics well and it should not be used in such cases. The moderate thermal shock resistance can lead to cracking when exposed to heat. If you were to pour boiling water into a cold bottle, you would find that the bottom would pop out. This also happens with aluminium oxide components, but at temperatures significantly higher than 100 °C.

The heat is well conducted by aluminum oxide, which is rather untypical for ceramic materials. Its thermal conductivity is 20–30 W/K m. It is significantly better than that of steels and is still almost half that of copper, which is a good conductor of heat (approx. 70 W/K m).

The thermal expansion of aluminium oxide is low, the coefficient of thermal expansion is about $7-10^{-6}$ K^{-1}. This is favorable because it means good dimensional stability of ceramic components at high temperatures.

The aluminum oxide has very good dielectric properties and thus it can insulate well. In other words, the oxide is a very poor conductor of electricity, technically speaking it has a high resistivity, which remains high even at high temperatures. The ability to insulate electrically well determines many applications of ceramics.

What else can be said interesting about the properties of aluminum oxide?

We can describe its chemical resistance as very good, it is extremely resistant to acids and alkalis. Only very aggressive solutions and melts such as hydrogen fluoride or hot concentrated nitric acid can attack the oxide and possibly dissolve it. Furthermore, we can use components made of high-purity aluminium oxide in many metal and glass melts, which extends its practical application.

Aluminium oxide ceramics belong to the bioinert materials. Even after a long exposure time, there is no reaction with body fluids, such as blood. Since grinding and polishing can also produce very smooth surfaces, aluminium oxide ceramics are used for hip joint implants where very low sliding friction is required. We will learn more about this in another chapter. The success of the surface treatment is strongly related to the grain size of the material. The finer the grain size of the aluminium oxide, the better it can be ground to a bright finish.

The aluminium oxide ceramic does not absorb water and is food safe. It is metallisable and can therefore be soldered.

This is a whole range of valuable properties that make many interesting applications of ceramics possible.

The Ceramic All-Rounder

27

Aluminium oxide is often called the "all-rounder" or the "workhorse" among the ceramic materials. We repeat: it is by far the most important ceramic material with a market share of almost 80%. The material owes these designations and its top position to the properties already mentioned, and also – this is also of great importance – to its low price or rather its good price/performance ratio. Despite the development of new high-performance ceramics, such as those based on silicon carbide and nitride, aluminium oxide ceramics continue to hold their market position and find a wide range of applications in technology.

In electrical engineering, various insulating parts are made from aluminium oxide ceramics. Spark plug insulators are manufactured almost exclusively from medium- and coarse-grained aluminas with a high aluminium oxide content (Fig. 27.1). The production of spark plugs was once the beginning of the triumphal march of ceramic materials. This already old application was also further developed. In 2010, a patent entitled "Spark plug with an insulator made of high-purity aluminium oxide ceramic" was approved in Germany.

Not only for spark plug insulators, but also for the production of high-voltage insulators, aluminium oxide with its excellent electrical insulation properties forms the ideal raw material basis.

In electronics, which so dominates our modern world, aluminium oxide ceramics are used as a substitute material for hybrid circuits. A striking symbol of electrical engineering are the famous "chips", silicon plates whose transistor function would be inconceivable without an aluminium oxide carrier.

In mechanical and plant engineering, wear protection parts and parts that are exposed to high temperatures are made of aluminium oxide. A good example is the textile industry. Many textile machines are characterised by some special features compared to general mechanical engineering: They work at considerable speeds in the production of textiles and they are sometimes subject to high stresses during operation due to temperature, pressure and chemicals. Ceramic materials, including our aluminium oxide, can withstand these stresses. Already at the beginning of this

Fig. 27.1   Spark plug with insulator made of aluminium oxide ceramic

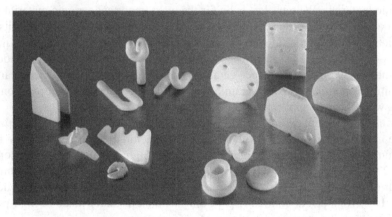

Fig. 27.2   Thread guides and formers made of aluminum oxide ceramics. (With the kind permission of the CeramTec company in Plochingen)

book we have become acquainted with a thread guide. In Fig. 27.2 we can again see these inconspicuous yet important parts, plus various formers. They are all used in the manufacture of fibres by melt spinning. Melt spinning is used for the production of fibres from meltable polymers such as polyamide or polyester. The high friction, surface wear and electrostatic forces that occur in this process can actually only be permanently endured by ceramic components.

In the chemical industry, aluminium oxide ceramics are mainly used for corrosion-resistant parts.

In high-temperature technology, combustion nozzles and protective tubes are made of this ceramic.

Cutting tools (in the form of indexable inserts) made of aluminium oxide ceramics are widely used in machining technology.

In medical technology, this ceramic is used very successfully for implants, which will be described in more detail in one of the next chapters.

The fact that the aluminium oxide is a powder makes it possible, in addition to sintering processing, to apply it as a coating. An interesting and important application. The coating of aluminium oxide is non-conductive and is therefore used for electrical insulation. Up to approx. 800 °C, this pure ceramic coating also offers good wear resistance. The ceramic coating has a structure that differs from the structure of solid ceramics. The most important feature: ceramic coatings have a process-related porosity of 1–5%, depending on the material. For this reason, ceramic coating is not recommended for corrosion protection.

Aluminium oxide has also a special industrial application as a regenerable desiccant.

A second book could easily be written about the technical applications of aluminium oxide. In this book, only two very different application examples are presented in more detail.

Hidden in a Scanning Electron Microscope

28

A small but important part made of aluminium oxide ceramics is found in every scanning electron microscope.

A scanning electron microscope (SEM) is an excellent device with which we can discover the world in a new way. The wealth of information we get with its help is almost immeasurable. An example: for a long time we did not know why water drops roll off a lotus leaf. Only an SEM image revealed the whole truth to us and today we can produce such water-repellent surfaces ourselves.

The function of the scanning electron microscope is based on the scanning of the sample surface by means of a finely focused electron beam. The electron beam is generated in an electron source, which is also very appropriately called an electrode gun and is electrically a cathode. Such a cathode looks very inconspicuous and yet is nothing ordinary. In the simplest case, it consists of a tungsten wire drawn out in the shape of a hairpin. The electrons are released from the cathode by applying heat. Current flows through the wire and heats it up to about 3000 °C. It begins to glow and emits electrons into its surroundings. These are first focused in a special cylinder (Wehnelt cylinder), then accelerated accordingly by the high voltage applied between the cathode and the anode. Now an electron beam is available. What such an inconspicuous SEM cathode looks like, we can see in Fig. 28.1; however, without tungsten wire, which unfortunately has already been burned.

This very hot tungsten cathode must be reliably held. Actually, only aluminium oxide ceramics with their high temperature resistance are suitable for this holder. And so the aluminium oxide even participates in scanning electron microscopy. We have also already been able to observe its microstructure on the basis of the SEM image (Fig. 24.2).

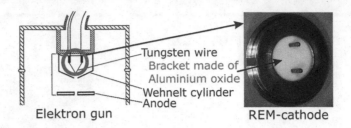

Tungsten wire
Bracket made of
Aluminium oxide
Wehnelt cylinder
Anode
Elektron gun   REM-cathode

Fig. 28.1 Used SEM cathode with holder made of aluminium oxide ceramic

Even Better in Compound 29

The whole is more than the sum of the parts – this is the principle of composite materials. Composites consist of two or more materials which, when combined, exhibit combinations of properties that none of the materials involved possesses on its own. This idea is also successfully applied to ceramic materials.

Poor properties of aluminium oxide ceramics, especially their brittleness, can be improved by using suitable particulate and/or fibrous materials. This modern method is called targeted material design. Composite materials created in this way are steadily gaining in importance.

Zirconium reinforced aluminium oxide is a very good example of a ceramic composite material. In English, this material is referred to as "zirconia toughened aluminium oxide" (ZTA ceramic). The material is produced by adding zirconia to the aluminium oxide matrix (Fig. 29.1). This results in a refinement of the microstructure compared to the pure aluminium oxide ceramic. The improved properties of the composite, particularly its higher toughness, are due to a crack-inhibiting mechanism of the zirconia, but this will not be discussed further here. Importantly, this more than doubles the resistance to crack propagation.

In addition to the zirconium oxide, this mixed ceramic contains other additives, which primarily include oxides of strontium, yttrium and chromium. The strontium oxide SrO is added to bring about the formation of the characteristic platelets, i.e. platelet-shaped crystals. These also cause an increase in toughness.

We recall that in the crystal lattice of aluminium oxide not all available sites are occupied by aluminum ions. These free places could be well exploited. A kind of doping can be carried out, which is sufficiently known from semiconductor technology, for example. Many research groups are eagerly working on modifications of aluminium oxide ceramics by specific impurities, if these additions may be called

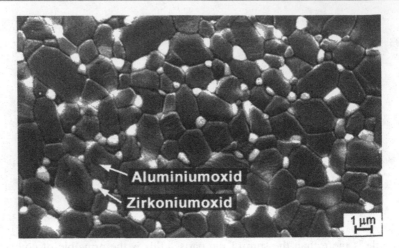

Fig. 29.1 Microstructure of a ZTA ceramic. (Source: www.keramverband.de)

so trivial. One of the rare earth elements, yttrium, seems to be particularly suitable here. Doping with yttrium atoms makes the ceramic much stronger, and the grains contract more tightly. The theoretical calculations revealed that the presence of the impurity atoms alters the local arrangement of the aluminum and oxygen atoms, thereby increasing the number of bonds. The resulting stronger internal cohesion makes the mixed ceramics more resistant to deformation at high temperatures. In technical terms, their creep resistance is significantly improved.

An important feature of the mixed ceramic of aluminium and zirconium oxide is its excellent biocompatibility. It represents the ideal material for further developments and future new applications in medical technology. We will find out more about this later.

A nice side effect of the zirconium-reinforced aluminium oxide ceramic (ZTA ceramic) is its color. It is pink/pink and thus quite characteristic. You could also say: This mixed ceramic simply looks good. And somehow it reminds of the beautiful red ruby, however, the ceramic is not transparent.

Another area of application for ceramic composites is machining technology. An interesting example of this is a mixed ceramic made of aluminium oxide and titanium hard materials (Fig. 29.2).

Its excellent wear resistance combined with outstanding hot hardness enables its use, for example, in the machining of hardened steels. It represents an economically interesting alternative to competing cutting materials made of cubic boron nitride CBN.

Fig. 29.2 Cutting materials made from the composite of aluminium oxide with titanium hard materials. (With the kind permission of CeramTec in Ebersbach)

The development of ceramic composites has actually only just begun. We can look forward to seeing which new materials will soon be ready for use.

Implants and No More Hip Joint Problems

30

Materials that are used in medical technology are particularly close to our hearts. We already know that aluminium oxide has good biocompatibility and is used for implants. Now we ask: For which implants then? Aluminium oxide ceramics are mainly used to make components for hip prostheses. The best known is the hip joint implant.

Many people suffer from hip pain: When walking, even lying down, the hip hurts and the mobility of the joint is limited. If the disease is advanced, the doctor advises an artificial hip joint. Today this is almost a standard method. The actually simple shape of the natural joint accommodates prosthetics: a spherical head engages in a cup-shaped socket. In the first prostheses, the ball was made of stainless steel and the socket of synthetic material, a special type of polyethylene. Unfortunately, the plastic wears out very quickly with this pairing.

The typical gait cycle results in about 20 million load changes in ten years, which causes abrasion. This is why it is essential for hip implants to have a high wear resistance material. Let's compare: An annual abrasion of a metal/polyethylene pairing is about 0.2 mm, while in a ceramic/ceramic pairing it can be reduced to only 0.001 mm, given suitable grades. The difference is simply enormous. Thus, the use of high-performance ceramics, especially aluminum oxide, has significantly improved hip prostheses. This happened in the early 1970s of the twentieth century.

Today, in addition to the pure aluminum oxide ceramic, the aluminum oxide-zirconium oxide mixed ceramic (ZTA ceramic) is also available for these implants. And today both components, the ball head and the socket, are made of ceramic materials. In Fig. 30.1 we can see the components of a hip joint prosthesis made of a special pink aluminium oxide mixed ceramic.

These ceramics are bioinert, i.e. they do not react with body fluids. They also have sufficient long-term mechanical strength. Aluminium oxide ceramics can be manufactured with an extremely smooth surface. When the ball head slides in the

B. Arnold, *Rubies and Implants*, https://doi.org/10.1007/978-3-662-66116-1_30

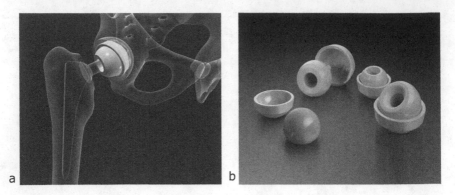

Fig. 30.1 Hip joint prosthesis made of aluminium oxide mixed ceramics. (**a**) General view, (**b**) Condylar heads and cups. (With the kind permission of the CeramTec company in Plochingen)

socket, there is almost no friction. At the same time, the low abrasion is also bioinert and the abrasion particles are very small. Thus, according to previous experience, inflammatory reactions do not occur. Further improvements can be achieved by special coatings of the ceramic parts.

Pink or rose-coloured implants in medical technology and red rubies in the science of healing stones. This is also part of the fascination of aluminium oxide.

News from Aluminium Oxide

At the end of a non-fiction book, it is fitting to write something new about the topic covered.

That natural types of aluminium oxide such as rubies and sapphires are transparent, we know, and we're not really surprised. But artificial, sintered aluminium oxide that is transparent? Is such a thing possible? Translucent aluminium oxide, made from nanoscale powders, has a high light transmission. By using an extremely fine-grained starting material (particle size smaller than 300 nm) and lowering the sintering temperature to 600 °C, it is even possible to produce transparent and pore-free ceramics. They can be used for novel halogen lamps, for scratch-resistant plates, for windows in special optical devices and also for invisible braces.

Aluminium oxide like mother of pearl? This also sounds kind of unlikely. And it's not about the appearance here, but about the properties. Mother-of-pearl not only fascinates with its iridescent sheen, but it also impresses with its high mechanical stability, thanks to a kind of "brick-and-mortar" structure. Nature once again proves to be an unbeatable inventor. Mother-of-pearl consists of 95% calcium carbonate, which is present in thin platelets stacked on top of each other. Between the platelets is a proteinaceous substance. This special and clever structure is apparently decisive for the good properties of the biominal. This is because nacre is about three thousand times as robust as pure calcium carbonate. Microscopic images show that a crack does not simply run through the material, but has to wind its way around the platelets. It follows a zigzag path, and its propagation is eventually stopped – a crack-stop effect. Catastrophic failure, as we know it from ceramics, does not occur.

Based on the model of mother-of-pearl, a new aluminium oxide ceramic was produced using a very special freezing process. Its mechanical properties amaze us. The material is ten times as strong as normal aluminium oxide ceramics, and

© The Author(s), under exclusive license to Springer-Verlag GmbH, DE, part of Springer Nature 2022
B. Arnold, *Rubies and Implants*, https://doi.org/10.1007/978-3-662-66116-1_31

at the same time extremely tough. A fracture does not pass through the ceramic, it is deflected several times until it finally ends. If the material is heated, it can withstand a temperature of at least 600 °C. Because common and less complex techniques are required for production, industrial implementation should be easy and soon possible.

This is the future, and aluminium oxide definitely owns it.

Further Reading

„Rubin, Saphir, Korund" extraLapis No. 15, Christian Weise Verlag, 1998

Walter Schumann „Edelseine und Schmucksteine", BLV Buchverlag, 2017

Ursula Wehrmeister, Tobias Häger „Edelsteine erkennen", Rühle-Diebener-Verlag, 2005

Horst H. Pohland „Aluminiumoxid-Herstellung, Eigenschaften, Einsatzgebiete", Die Bibliothek der Technik, Band 176, verlag moderne industrie, 1998

Michael F. Ashby, David R.H. Jones „Werkstoffe 2: Metalle, Keramiken und Gläser, Kunststoffe und Verbundwerkstoffe", Spektrum Akademischer Verlag, 3. Auflage, 2007

Martin Heinrich Klaproth „Beiträge zur chemischen Kenntniss der Mineralkörper", 1. Band, Posen, Berlin 1795

Dagmar Hülsenberg „Keramik – Wie ein alter Werkstoff hochmodern wird", Springer Vieweg, 2014

Index

Printed in the United States
by Baker & Taylor Publisher Services

Printed in the United States
by Baker & Taylor Publisher Services